Chemical Process Safety

Chemical Process Safety

Editor

Aarti Kashyap

Chemical Process Safety

Edited by **Aarti Kashyap**

Printed in 2017

ISBN: 978-1-68117-343-6

Library of Congress Control Number: 2015939255

© 2016 by
SCITUS Academics LLC,
616, Corporate Way, Suite 2, 4766,
Valley Cottage, NY 10989

www.scitusacademics.com

This book contains information obtained from highly regarded resources. Copyright for individual articles remains with the authors as indicated. All chapters are distributed under the terms of the Creative Commons Attribution License, which permits unrestricted use, distribution, and reproduction in any medium, provided the original author and source are credited.

Notice

Reasonable efforts have been made to publish reliable data and views articulated in the chapters are those of the individual contributors, and not necessarily those of the editors or publishers. Editors or publishers are not responsible for the accuracy of the information in the published chapters or consequences of their use. The publisher believes no responsibility for any damage or grievance to the persons or property arising out of the use of any materials, instructions, methods or thoughts in the book. The editors and the publisher have attempted to trace the copyright holders of all material reproduced in this publication and apologize to copyright holders if permission has not been obtained. If any copyright holder has not been acknowledged, please write to us so we may rectify.

Contents

Preface vii

Chapter 1   Economic-Oriented Stochastic Optimization in Advanced Process Control of Chemical Processes 1

László Dobos, András Király, and János Abonyi

Chapter 2   Sustainability Assessment of Chemical Processes: Evaluation of Three Synthesis Routes of DMC 27

Paula Saavalainen, Satish Kabra, Esa Turpeinen, Kati Oravisjärvi, Ganapati D. Yadav, Riitta L. Keiski, and Eva Pongrácz

Chapter 3   Improved Safety for Automotive Lithium Batteries: An Innovative Approach to include an Emergency Cooling Element 53

Peter Kritzer, Harry Döring, and Brita Emermacher

Chapter 4   Process for the Obtention of Coumaric Acid from Coumarin: Analysis of the Reaction Conditions 73

Néstor N. López-Castillo, Alma D. Rojas Rodríguez, Brenda M. Porta, and M. Javier Cruz-Gómez

Chapter 5 CO_2 Utilization: A Process Systems Engineering Vision 87

Ofélia de Queiroz F. Araújo, José Luiz de Medeiros and Rita Maria B. Alves

Chapter 6   A Future Perspective on the Role of Industrial Biotechnology for Chemicals Production 157

John M. Woodley, Michael Breuer, and Daniel Mink

Chapter 7   Safety Analysis Approach Based on Thermodynamic and Chemical Reactions Modelling 179

Taha Benikhlef, Djamel Benazzouz, Smail Adjerid, and Kazimierz Lebecki

Chapter 8	**Conventional and Dynamic Safety Analysis: Comparison on a Chemical Batch Reactor**..215
	L. Podofillini and V.N. Dang

Citations..251

Index..255

Preface

The primary objective of this book is to encapsulate the important technical fundamentals of chemical process safety. The emphasis on the fundamentals will help the student and practicing scientist to understand the concepts and apply them accordingly. This application requires a significant quantity of fundamental knowledge and technology. Chemical Process Safety is designed to enhance the process of teaching and applying the fundamentals of chemical process safety. It is appropriate for an industrial reference, a senior-level undergraduate course, or a graduate course in chemical process safety. It can be used by anyone interested in improving chemical process safety, including chemical and mechanical engineers and chemists.

Editor

Economic-Oriented Stochastic Optimization in Advanced Process Control of Chemical Processes

László Dobos, András Király, and János Abonyi

Department of Process Engineering, University of Pannonia, Egyetem Street 10, 8200 Veszprém, Hungary

ABSTRACT

Finding the optimal operating region of chemical processes is an inevitable step toward improving economic performance. Usually the optimal operating region is situated close to process constraints related to product quality or process safety requirements. Higher profit can be realized only by assuring a relatively low frequency of violation of these constraints. A multilevel stochastic optimization framework is proposed to determine the optimal setpoint values of control loops with respect to predetermined risk levels, uncertainties, and costs of violation of process constraints. The proposed framework is realized as direct search-type optimization of Monte-Carlo simulation of the controlled process. The concept is illustrated throughout by a well-known benchmark problem related to the control of a linear dynamical system and the model predictive control of a more complex nonlinear polymerization process.

INTRODUCTION

Due to the dynamic and significant changes of the economic environment performance assessment of process control is highlighted area of chemical engineering [1]. The aim of this paper is to develop an optimization framework designed to determine optimal operating regimes of chemical processes by taking process constraints, desired maximum number (frequency) of constraint violations, and process uncertainties into consideration.

Variance in the closed control loop caused by unmeasured disturbances and badly designed controllers might cause variations in the product quality. In case of increasing variance of process variables the probability and frequency of violation of quality requirements are increasing that might lead to the increase of the amount of less valuable offset products. Typical examples when reduced number of violations of the predetermined process constraints are acceptable can be found in the field of statistical process control (SPC [2]). In Statistical Process Control statistical tools are applied to monitor the performance of the production process and detect significant deviations that may later result in offset products.

In this paper a more sophisticated model-based approach is followed. Modern process analysis, monitoring, control, and optimization tools are mainly based on some kind of process model. It is obvious to utilize these process models also in the economic assessment and optimization. Usually the output of cost-benefit analysis is cost reduction or profit increment expressed by a cost function. These functions incorporate the costs of the operation, raw materials, current prices of products [3], and risks of malfunctions. In our economic-oriented optimization strategy the aim is to find steady state operation points (controller set points) where profit might be realized. This task is fulfilled at the supervisory control level [4].

The general approach for economic performance evaluation comprises the following steps: reduce the variance of the controlled variable and shift the set points (process mean) closer to the operation limits [5] without increasing the frequency of the violation. This operation is referred to as the improved control [6]. The variance reduction might mean to retune the existing controllers, or, in more radical cases, change the whole control strategy. The model-based predictive controllers (MPCs [7]) is highly applicable for variance-reduction purposes. Application of MPCs in the operative control level results in a multilayer optimization problem, since an MPC also minimizes its objective function. In this approach the upper layer is the supervisory control level which is responsible for economically optimal operation, and the lower layer is for variance reduction.

To handle uncertainty and effects of measurement noise in this paper a novel Monte-Carlo simulation-based approach is proposed. Monte-Carlo simulation is frequently applied in various areas [8]. This tool has also proven its efficiency in risk-related optimization of chemical processes; for example, it is applied in optimizing maintenance strategies of operating processes [9]. There is a common characteristic in these solutions: the stochastic nature of the studied system has to be modeled. In the applied methodology this simulation is related to the modeling of the unmeasured disturbances of the control loops. To handle this random effect, Monte-Carlo simulation is applied with the characterized noise. An economic cost function is calculated in every case to measure the economic efficiency of the process. Integrating this benefit analysis tool into the mesh adaptive direct search optimization algorithm—where the task is to find the most beneficial steady state operation point—resulted in the proposed economic-oriented optimization framework. In the proposed multilayer optimization framework, the application of gradient-based methodologies for maximizing the economic throughput is not possible, thanks to the stochastic characteristics caused by the closed-loop variance. That is why the utilization of direct search methods is necessary. Mesh Adaptive Direct Search (MADS) [10] class of algorithms is a relatively new set of direct search methods for nonlinear optimization; that is, these algorithms are capable of calculate the extremums of a nonsmooth functions, like our economic objective function. Since the steady state operation points are mainly determined by the variance of the controlled variables, incorporating this effect into the model is inevitable. The created optimization framework functions as an industrial Advanced Process Control system, [3, 11].

The paper is organized as follows: in Section 2 the economic cost function-based multilayer optimization framework is introduced. In Section 3 the applied methodology is explained in detail. In Section 4, the efficiency of the proposed methodology is illustrated throughout a linear benchmark control problem and a Model Predictive Controlled (MPC) highly non-linear technology. In both cases an economic performance measure has been formalized as a basis for optimizing the set point signal. As base case of the benchmark example the process is controlled with a PI controller. To reduce the closed-loop variance caused by unmeasured disturbance a linear MPC is installed to replace the PI controller. With the reduction of the variance the set point of the controller can be moved closer to the process constraints which yields higher economic performance. Such economic-oriented optimization is carried out at two different risk levels. As a second example a non-linear process controlled by a linear MPC is considered, since this combination is widely applied in chemical process industry. In this case study the process variance is caused by an unmeasured disturbance, model mismatch, and noise added to the controlled variable. In this example the effect unmeasured disturbance with different amplitude is examined in detail. These examples show the realistic benefits of the proposed methodology.

ECONOMIC COST FUNCTION-BASED MULTILAYER OPTIMIZATION

The proposed framework is rather similar to Advanced Process Control (APC) systems applied in the chemical process industry, [3]. The scheme of this multilayer optimization problem is depicted in Figure 1. The main aspects and tasks that have to be taken into consideration in the different optimization levels will be introduced in the following subsections.

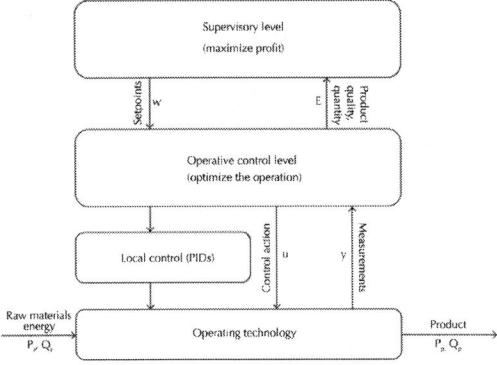

Figure 1: The layers of an economic optimization of an operating technology.

The Supervisory Control Level

The main task in the supervisor level is to maximize the economic throughput with varying the steady state set point signal. In general the economically optimal set point is close to the operation limits of process. That is why the reduction of the closed-loop variance is necessary. Thanks to process variance—caused by disturbances, noise, and so forth—there is a risk of process constraint violation which has to be taken when the new set point is determined. The essence of this economic optimization approach is depicted in Figure 2.

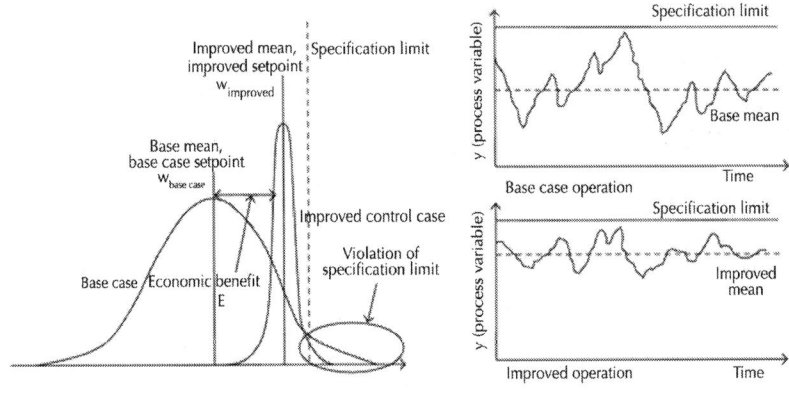

Figure 2: Approach to economic benefit estimation with variance reduction.

The aim of the economic-oriented process optimization framework can be formulated as maximizing a cost function. Such cost function mostly includes the cost of the actual operation, raw materials, and the value of the product as the following:

$$\max_{w} E = \sum_{i=1}^{N} P_i \cdot Q_i - \sum_{r=1}^{M} P_r \cdot Q_r, \tag{1}$$

where P_p, P_r and Q_p, Q_r are the prices and the quantities of the products and raw materials, respectively. In the optimization problem $w = [w_1, \ldots, w_p]^T$ represents the setpoints of local controllers of the operative control level; p is the number of the controlled variables, denoted with y_i. The task of the operative control level can be summarized as y_i should be as close to w_i as possible.

Continuous economic improvement of process control is about to find the setpoint values of control loops with the possible highest economic performance. In steady state operation the value of (1) is often increasing by shifting the steady state operation point closer to the constraints of the process. To reach this goal, the reduction in variance of the key process variables is necessary with, for example, retuning the controller or even redesigning the existing control strategy. As Figure 2 illustrates, when the variance of the key process variables is reduced extra profit can be realized as the difference of the economic potential in the old and revised steady state operation points.

The cost function must be optimized with respect to the process constraints to ensure the required process safety and product quality. Constraints defined on the process variables can be expressed as follows:

$$y_{i,\min} \leq y_i \leq y_{i,\max}, \quad i = 1, \ldots, p,$$

$$u_{j,\min} \leq u_j \leq u_{j,\max}, \quad j = 1, \ldots, m, \tag{2}$$

where $y_{i,min}$ and $y_{i,max}$ are the lower and upper bounds of output variables, $u_{j,min}$ and $u_{j,max}$ are the lower and upper bounds of input variables, and p and m are the number of output and input variables, respectively. Thanks to uncertainties, such as process variation and disturbance, the probability of violating the predetermined process constraints is increasing by getting closer to them. A reasonable approach to handle the uncertainties in the constraints is to cast the the problem in term of the probability of constraint violation, which is the approach to be implemented in this paper

The probability constraints can be expressed as

$$\Pr\{y_{i,\min} \le y_i \le y_{i,\max} \ i = 1,\ldots,p\} \ge 1 - \alpha, \qquad (3)$$

Or

$$\Pr\{y_{i,\min} \le y_i \le y_{i,\max}\} \ge 1 - \alpha_i \quad i = 1,\ldots,p, \qquad (4)$$

where Pr{·} is the operator of probability and α is the specified probabilistic violation level (demonstrated in Figure 2). The formulation of probability constraints means that satisfying process constraint of y_i is not required by 100% probability but a certain confidence level, 1 − α. Inequality (3) represents a so-called Joint Probabilistic Constraint (JPC) problem, which means that all process variables must be kept in the defined operation regime with maximum probability of violation of α. Inequality (4) is called is Individual Probabilistic Constraint problem (IPC) ([6]), where each process variable has a specified confidence level, 1 − α_i, to be satisfied. In this paper the second approach has been adopted.

The final goal in multilayer optimization is to maximize the accessible profit, determined by the cost function mentioned before in (1), by finding the optimal steady state operation point with respect to the process constraints, (2). Thanks to process variation a reasonable risk level has to be taken by defining the probability of process constrain violation (formalized as α). 1 − α means a confidence level, which is a non-linear constraint in the economic optimization. This optimization problem represents the supervisory control layer.

The Operative Control Level

In optimization and control of complex production processes, the role of Model Predictive Controllers (MPCs) is increasing. The more and more widespread application is reasonable, thanks to the good variance reduction ability.

MPC is a model-based control algorithm where models are used to predict the behavior of process outputs of a dynamical system with respect to changes in the process inputs. The MPC uses the models and current plant measurements to calculate future moves in the manipulated variables, which will result in operation that honors all input and output variables' constraints (see (2)).

Predictive control uses the receding horizon principle. This means that after the computation of the optimal control sequence, only the first control action will be implemented; subsequently, the horizon shifted one sample and the optimization is restarted with new information about the measurements. That is the reason why the MPCs do not optimize the operation on the time horizon of the whole steady state operation, but consider just the horizon, implemented in the controller, and solve the optimization problem iteratively. With the help of Figure 3 the essence of the model predictive control is easily understandable.

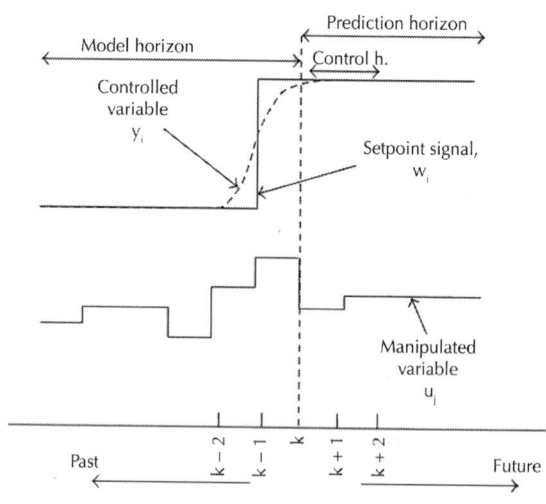

Figure 3: Illustration of essence of MPC in control of Y_i by manipulating u_j.

Utilizing this control strategy in the operative control level the previously proposed multilayer optimization framework can be resulted (depicted in Figure 1), thanks to the control rule of MPCs, expressed as:

$$\min_{\Delta u_{(k+j)}} \sum_{i=1}^{p} \sum_{j=1}^{H_p} \left(w_{i,k+j} - y_{i,k+j}\right)^2 + \sum_{i=1}^{m} \lambda_i \sum_{j=1}^{H_c} \Delta u_{i,k+j-1}^2, \quad (5)$$

where p and m are the numbers of the controlled variables and manipulated variables, respectively. In MPC control strategy the different number of controlled and manipulated variables is acceptable, since the interconnection between the different manipulated and controlled is considered in the process model, which is applied in the MPC. The tuning parameters of the controller are as follows: H_p and H^c are the length of prediction and control horizo, λ is the a factor for punishing the change of the control signal. $\Delta\upsilon$ is the variation of the manipulated variable at a given time, which is calculated during the optimization method on H_c control horizon.

MPC formulates an objective function which is used to find the optimal input sequence to eliminate the difference of the controlled variable and the set point in the prediction horizon. Since this objective function, (5), is designed to ensure smooth and stable operation it does not directly reflect the economic performance of the technology (formalized in (1)). Additionally this cost function does not count to the risk of violating the process constraints caused by unmeasured disturbances which appears as closed-loop variance (see (3)-(4)).

Constraint violations have also to be taken into account during economic performance optimization. This is the reason why this paper suggests the application of Monte Carlo simulation of the augmented process model and the model of the operative control level (see Figure 1). The result of the Monte Carlo simulation is an aggregated economic performance (e.g., mean of the economic performance of the individual runs). Due to the stochastic nature of the optimized system the gradient of this aggregated economic cost function is difficult to calculate. Hence the optimization algorithm should be gradient free-yet computationally very effective. To meet this requirement the application of the advanced Mesh Adaptive Direct Search methodology is proposed. In the following section the multilayer optimization framework and its two main building blocks—Monte Carlo simulation and Mesh Adaptive Direct Search methodology—are going to be introduced in detail.

STOCHASTIC MODELING ECONOMIC BENEFIT MAXIMIZATION WITH DIRECT SEARCH METHODOLOGY

Taking process variance into account the previously proposed economic-oriented objective function, (1), becomes a stochastic characteristic. To handle uncertainties Monte Carlo simulation is applied. The Monte Carlo method is applied frequently in solution of stochastic optimization problems, for example, in stochastic linear programming [12, 13]. Kjellstrom [14] was the first to use Monte Carlo estimators for the iterative improvement of convergence behavior in nonlinear stochastic optimization.

Due to the stochastic characteristics caused by the closed-loop variance the application of gradient-based methodologies for maximizing the economic throughput is not efficient. Integrating the simulation-based economic performance assessment methodology into a direct search optimization algorithm an effective optimization framework is obtained. Mesh Adaptive Direct Search (MADS) [10] class of algorithms is a relatively new set of direct search methods for nonlinear optimization; that is, these algorithms are capable of calculating the extremums a nonsmooth objective functions, like our economic objective function.

Our methodology is stated as follows:
- economic performance assessment of the considered steady state operation point. It means applying a set point (w) and calculating the value of the economic cost function, (1), with respect to the process constraints, (2), and the value of the probability of constraint violation (see (3)-(4)). Because of considering the process variance as random phenomena, Monte Carlo simulation with multiple runs of augmented process simulator is applied to aggregate the effect of the random variances in a final economic cost function;

- integrate the economic performance evaluation tool into the MADS optimization algorithm to find the economically optimal steady state operation point. The previously applied economic cost function, (2), has to be maximized with respect to the proposed constraints with varying setpoint signal (w). This algorithm can handle constraint limits of process variables, the certain confidence levels to violate these limits.

Using the methodology discussed above the optimization process is capable to isolate and handle all the disturbances technology has, whose nature is constant in time; thus it can be characterized statistically. These uncertainties are time homogeneous and static-time disturbances, like measurement noise or model error. In the following section the application way of Monte Carlo simulation and MADS optimization algorithm is introduced briefly.

Monte Carlo Simulation

Monte Carlo Simulation (MCS) methods are highly applied in the mathematical modeling problems where some kind of stochastic phenomena must be handled. In the proposed multilayer optimization framework process variance caused by unmeasured disturbances is considered. The Monte Carlo simulation consists of the following steps.

- Define the domain of possible inputs.
- Generate inputs from this domain randomly using a specified probability distribution.
- Execute deterministic computation using the inputs.
- Aggregate the results of the computations into the final result.

In engineering practice normal distribution is considered as an adequate assumption for characterizing uncertainties. At the modeling of the considered process the following steps are followed: at first the mathematical model of the process is created. Then noise and unmeasured disturbances of the control loops are characterized and random signals related to the real process variance are added to the corresponding input and output variables. The value of the economic objective function, (1), is calculated by aggregating the results of the individual Monte Carlo runs into a statistical economic performance. Since complex production processes are mostly characterized by non-linear process models the economic assessment and optimization needs an optimization algorithm which is able to handle the non-linear cost functions and constraints, (2), (3), and (4).

The Mesh Adaptive Direct Search Methodology

Since the calculation of the gradient of the economic objective function with respect to the steady state operation points is highly computational demanding and due to the Monte Carlo simulation the economic cost function is nonsmooth the application of gradient free optimization method is needed. Mesh Adaptive Direct Search (MADS) [10] is a relatively new set of direct search methods for nonlinear optimization. This algorithm is capable of minimizing a nonsmooth function, like our economic cost function (1) under the proposed constraints in (2) and (4). According to [10, 15], MADS can be interpreted as a generalization of Generalized Pattern Search (GPS) [16] algorithms, with the restriction to finitely many pool direction removed.

MADS is an iterative algorithm, where at each iteration a finite number of test points are generated. At the beginning of an iteration, the infeasible test points are filtered (discarded); that is, infinite objective value is assigned to it ($f(x) = +\infty$). Thereafter the feasible test points are evaluated by the objective function and compared with the current best objective function value found so far. Each of these test points lies on the current mesh, which is constructed from a finite set of n_D directions $D \in \square^n$ and scaled by the mesh size parameter $\Delta_k^m \in \square^n$. If we find a point with lower objective value than the current best one, this test point is a so-called improved mesh point and the iteration is a successful iteration.

Each iteration consist of two steps, the so-called SEARCH step and POLL step. SEARCH step can return any point of the underlying mesh; it is trying to find an unfiltered point. If it fails to generate an improved mesh point, then the second step, the POLL is invoked. POLL step consists of a local exploration around the current best solution, and the test points are generated in some directions scaled by the mesh size parameter. MADSs are novel in the number of usable directions, since In GPS, POLL directions belong to a finite set, while POLL direction in MADS belongs to a much larger set; in fact if the iteration number k goes to infinity, the union of the normalized POLL directions over all k becomes dense in the unit sphere. According to [10], this algorithmic construction allows stronger convergence. Another important difference between MADS and GPS is the so-called poll size parameter, Δ_k^p. This parameter determines the size of the frame where the POLL step can operate. In case of GPS, mesh size and poll size are equal ($\Delta_k^m = \Delta_k^p$), while in MADS these two parameters can differ. This difference is depicted in Figure 4. Additional pieces of information like convergence analysis or practical implementations can be found in [15].

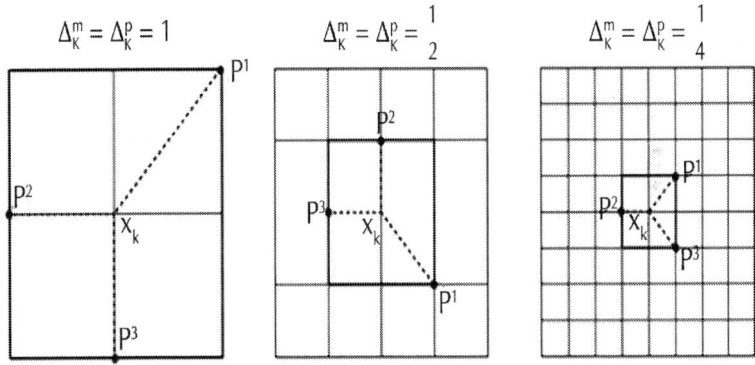

Figure 4: Example of GPS frames (a) and MADS frames (b) $P_k = \{x_k + \Delta\mu_k d : d \in D_k\} = \{p^1, p^2, p^3\}$ for different values of $\Delta_k^m = \Delta_k^p$. In all six figures, the mesh Mk is the intersection of all lines.

In the economic-oriented multilayer optimization framework (see Figure 1) MADS is applied in the supervisory level to maximize the economic performance formalized as (1). The optimization problem is solved with respect to the the process constraints, (2), and the value of the probability of constraint violation, see (3)-(4), with varying setpoint signal (w). Since MADS needs a reduced number of runs of the augmented process simulator, the optimal value of the setpoint signals can be quickly obtained. The low number of iteration during optimization is necessary, since Monte Carlo simulation of the operative control level (augmented process simulator) is applied, which is highly computation demanding process.

In the following section the effective application of the proposed framework is going to be examined throughout the case studies of a benchmark, linear process, and an MPC controlled highly non-linear technology.

APPLICATION EXAMPLES

In this section, two application examples are presented to demonstrate the applicability of the proposed framework for enhancing the economic benefit of the operating technologies. The calculations for both examples are based on closed-loop data, generated using Matlab-Simulink. The uncertainties are presented in the examples as noise superimposed to inputs and outputs.

A SISO Process

Consider a SISO process, characterized by G_p shown in Figure 5 subject to disturbance dynamics G_d described by

$$y_k = G_p u_k + G_d \alpha_k = \frac{0.6299 z^{-1}}{1 - 0.8899 z^{-1}} u_{k-2}$$

$$+ \frac{1 - 0.8 z^{-1}}{1 - 0.8899 z^{-1}} \alpha_k \quad k = 1 \ldots q, \tag{6}$$

where α_k is a normally distributed white noise sequence of mean 0 and variance 1. q signs the last time step of the considered simulation. The objective in the supervisory control level is to maximize the output ($\bar{y}$—mean of the output on the considered time horizon) with respect to the process constraints. The optimization problem can be formalized as

$$\max_{w} 2\bar{y} \tag{7}$$

subject to

$$-10 \leq y_k \leq 10$$

$$-5 \leq u_k \leq 5$$

$$k = 1\ldots q.$$

(8)

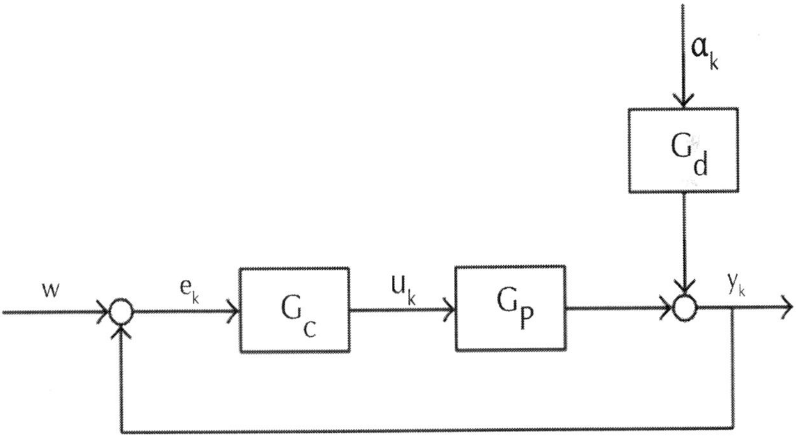

Figure 5: Block diagram of the SISO closed-loop system.

As base case a PI controller is designed. The controller parameters are $K_c = 1.926$, $T_I = 0.6$. As previously shown, specifying the probability of not violating the constraint defined on the output variable defines a non-linear constraint for the optimization problem. During the presented studies this confidence level is assigned as 95% and 90%. In the literature [6] the same SISO process is utilized with the same probabilities. The means of the output are $\bar{y} = 1.49$ and $\bar{y} = 2.72$ confidence level of 95% and 90%, respectively. The output data in 95% confidence level is depicted in Figure 6.

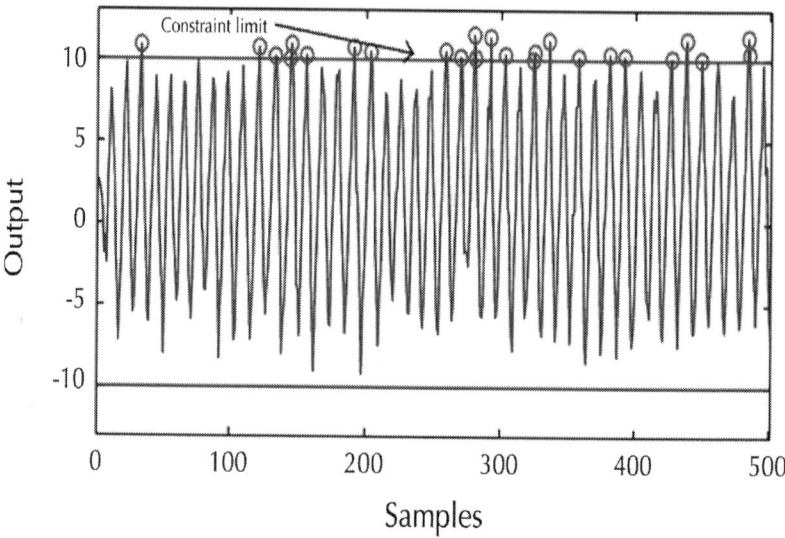

Figure 6: Base case operation with probability constraint level of 95%.

Since MPC is highly applicable for variance reduction purposes the PI controller has been replaced with a linear Dynamic Matrix Controller (DMC) [17]. DMC applies the linear convolution model of the process for predicting the effects of the considered manipulated variable sequence. With the application of DMC lower variance ($\bar{\sigma}_{PI}$ = 5.3 in contrast to $\bar{\sigma}_{DMC}$ = 1.05) and higher economic benefit might be expected. The control rule of the DMC has been proposed in (5). The tuning parameters of the applied DMC are H_m = 50, H_p = 20, and H_c = 10. The value of λ is chosen as 1000. By applying the previously proposed multilayer optimization framework significant improvement could be experienced in the economic performance (the number of Monte Carlo iterations was set to 100). The means of the output are $\bar{y}$ = 8.28 and $\bar{y}$ = 8.68 at confidence level of 95% and 90% respectively. The outputs in the improved operation are depicted in Figures 7 and 8. It can be clearly stated when the confidence level decreases the frequency of constraint violation increases.

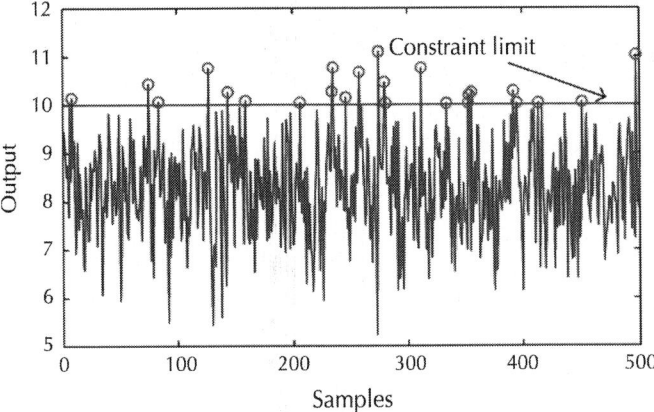

Figure 7: Improved case, optimal operation with probability constraint level of 95%.

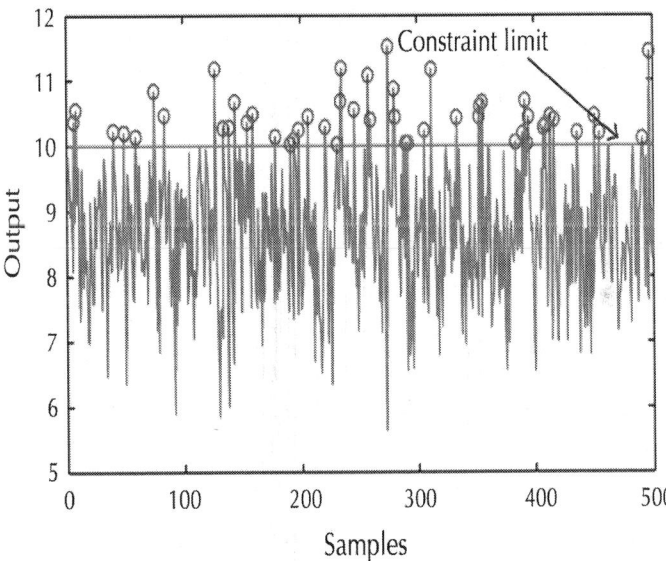

Figure 8: Improved case, optimal operation with probability constraint level of 90%.

The number of individual economic performance evaluations in the Monte Carlo-simulation has been set to 100. There has been an attempt to apply quadratic programming as optimization algorithm (utilizing Matlab, Optimization Toolbox), but the computational demand was extremely high, almost one hour even in this simple example. By applying MADS, the computation demand has been significantly decreased into 5 minutes. In both cases the initial setpoint for the optimizer was set equal to the upper constraint of the output variable, $w_0=10$.

In Figure 9 achievable economic benefit is depicted. Thanks to the replacement of the PI controller with the DMC the variance in the closed-loop could be reduced. Utilizing the previously introduced Monte Carlo simulation based optimization methodology new steady state operation points have been determined with multiplied economic performance with respect to the defined confidence level.

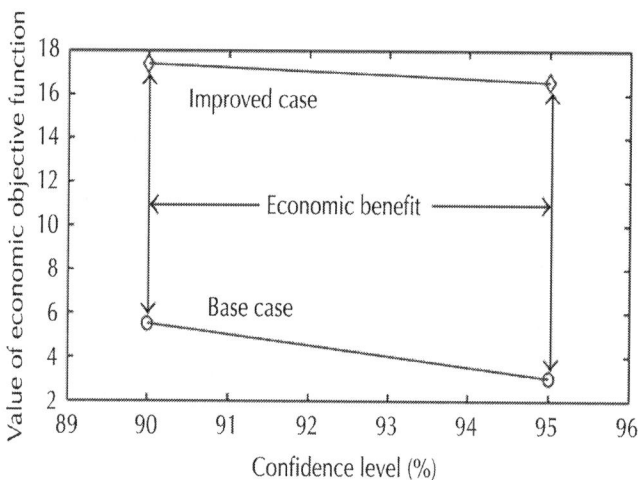

Figure 9: The economic performance in the base and improved case at different confidence levels.

The Polymerization Process

The process under consideration is a polymerization process controlled by a linear MPC, the previously mentioned DMC, [17]. The controlled system possesses all those difficulties which exist in an operating polymerization process.

Process Description

The reactor which has been studied is a CSTR where a free radical polymerization reaction of methyl-metacrylate is considered using azobisisobutyronitrile (AIBN) as initiator, and toluene as solvent. The aim of the process is to produce different kinds of product grades. The number average molecular weight is used for qualifying the product and process state. The polymerization process can be described by the following model equations [18]:

$$\frac{dC_m}{dt} = -(k_p + k_{fm})C_m P_0 + \frac{F(C_{min} - C_m)}{V}$$

$$\frac{dC_I}{dt} = -k_I C_I + \frac{F_I C_{Iin} - FC_I}{V}$$

$$\frac{dD_0}{dt} = (0.5k_{tc} + k_{td})P_0^2 + k_{fm}C_m P_0 - \frac{FD_0}{V}$$

$$\frac{dD_1}{dt} = M_m(k_p + k_{fm})C_m P_0 - \frac{FD_1}{V}, \qquad (9)$$

Where

$$P_0 = \sqrt{\frac{2f^* C_I k_I}{k_{td} + k_{tc}}} \qquad (10)$$

The notation for the equations can be seen in Table 1. The number average molecular weight (NAMW) is defined by the ratio of D_1/D_0. By assuming an isotherm operation model the process model consists of four states, represented by four differential equation (9) [19]. For the integration, the MATLAB's built-in ode45 function has been used, which is based on an explicit Runge-Kutta (4,5) formula. During simulations $T_s = 0.03h$ is applied as sample time.

Table 1: Design parameters for MMA polymerization reactor

F	1.0 m3/h
Cmin	6.4678 kmol/m3
Clin	8 kmol/m3
V	0.1 m3
Mm	100.12 Kg/Kmol
F*	0.58
R	8.314 Kj/Kmol K
Kp	2.4952·106 m3/kmol h
KI	1.0224·10−1 1/h
Kfm	2.4522·103 m3/kmol h
Ktc	1.3281·1010 m3/kmol h
Ktd	1.0930·1011 m3/kmol h

MPC Controller Strategy and the Economic Performance Assessment

The qualification of the product and process operation is based on the number average molecular weight. Thanks to the non-linear model equations the development economic performance turns into a highly non-linear optimization problem.

The control objective on the supervisory control level is to maximize the economic performance of the process. The objective function is formalized as

$$\max_{w_{NAMW}} E = P_{onspec} \cdot Q_{onspec} - P_{offspec} \cdot Q_{offspec},$$

(11)

where P_{onspec} (with the value of 10000) and $P_{offspec}$ (with the value of 3500) are the prices of the polymer product which fulfill/not fulfill the product specifications; w_{NAMW} indicates the steady state setpoint. Q is the quantity of the polymer product, calculated with the following expression:

$$Q_{polymer} = F \cdot D_1. \tag{12}$$

During the economic-oriented optimization, the following process constraints have to be considered:

$$24000 \leq NAMW \leq 26000. \tag{13}$$

An important characteristic of the process is the increasing of product quantity when shifting the steady state operation closer to the lower limit, hence the optimal steady state operation point is expected near to the lower limit. The maximum probability of violating the process constraints is 1%, so the mentioned confidence level is 99%. The number of Monte Carlo simulations is 100, similarly to the previous case. The closed-loop variance of the process is caused by the noise added to the inlet monomer flowrate (F) with means of 0 and $\bar{\sigma} = 0.014$. Other source of the closed-loop variance is the noise added to the controlled variable with the mean of 0 and $\bar{\sigma} = 143$.

On the operative control level the previously proposed DMC is designed. The manipulated variable in the control strategy of the reactor is the initiator inlet flow rate. The tuning parameters of the applied DMC are $H_m = 30$, $H_p = 3$, and $H_c = 3$. The value of λ is chosen as 4.10^{12}.

As base case the safest steady state operation point has been chosen which is in the middle of the specified operation range ($w_{NAMW} = 25000$). As the result of the economic performance optimization the optimal steady state setpoint is $w_{NAMW} = 24300$. Thanks to this setpoint modification the quantity of the produced polymer has been increased with 5% and throughout this the economic performance also increased with 5%. The result of the closed-loop simulation with the optimal setpoint is depicted in Figure 10.

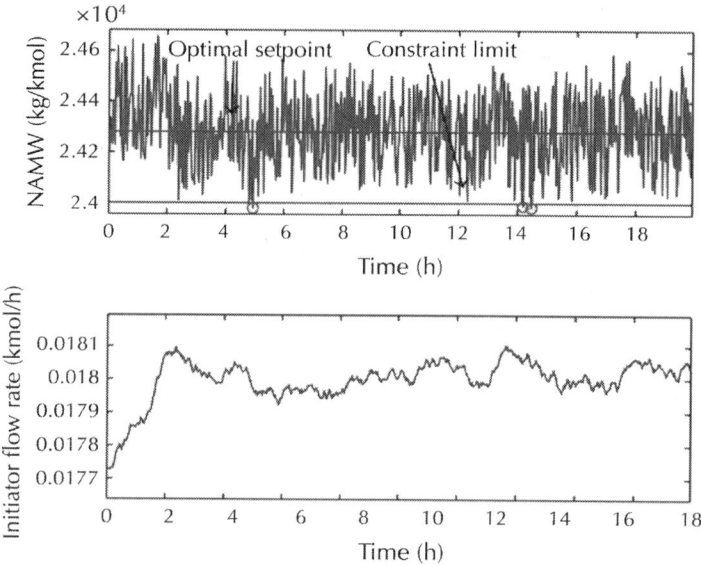

Figure 10: Improved case, optimal steady state operation point of PMMA reactor.

The number of individual economic performance evaluations in the Monte Carlo simulation has been set to 100. In this case study there has been an attempt to apply quadratic programming as optimization algorithm. The same result has been obtained with quadratic programming but the computational demand was extremely high, almost 10 hour. By applying MADS, the computation demand has been decreased into 1 hour. As it can be seen, the quadratic programming might be applicable but its computation demand is exaggerated. The initial setpoint for the optimizer was set equal to the lower constraint of the output variable, $w_0 = 24000$.

As Figure 10 shows that the frequency of constraint violation is conspicuously low, the process constraint has been violated only three times. It means 99.55% probability of not violating the limits, in contrast to the previously determined confidence level, which was 99%. It may happen since the off-specification product (products which do not fulfill the requirements) means extra outgoings in the economic objective function. Accordingly it is not worth to produce even just 1% off-specification product; however, this amount can be accepted technologically.

If the circumstances of the steady state operation change, there will be the need of redetermining the optimal operation point. This case is considered when the the deviation of the noise on the monomer inlet flow rate has been increased to $\bar{\sigma}=0.03$. Since the variation of the closed loop is increased, the optimal setpoint is determined further from the specification limit, $w_{NAMW}=24400$. This way the economic performance is decreased compared to the previous case with 0.5%.

The results confirmed the assumption that the economically optimal operation is close to the process constraints. However, 99% was set as confidence level of limit violation; the way of formulating the economic cost function does not allow such a low quantity of off-specification product, since it causes extra outgoings during operation.

CONCLUSIONS

In this paper an economic-oriented optimization framework has been introduced to determine optimal operation regimes of complex chemical process systems. Situations where the economically optimal steady state operation point is close to the technological limits of the operation have been studied. Due to process variance caused by unmeasured disturbances and measurement noise the determination of the optimal values of the controller setpoints is rather difficult since shifting the operation point closer to the process limits results in risk of violation of constraints related to process safety and product quality requirements. By formulating an economic objective function the performance and the risk level of the operation can be quantitatively evaluated. Monte Carlo simulation is applied with multiple runs of process model (augmented with the model of the control system) with economic performance assessment to handle the stochastic phenomena of process variance. Integrating the Monte Carlo simulation-based economic performance assessment tool into the Mesh Adaptive Direct Search (MADS) methodology can take process constraints and desired risk/confidence levels into consideration. Since MADS is the one of the recent gradient-free optimization methodology with high efficiency its application in Monte-Carlo simulation is much more effective than the classical gradient-based quadratic programming (SQP) by using MADS the time demand of optimization can be shortened to one-tenth of SQP.

The efficiency of the proposed framework is demonstrated throughout benchmark examples. In the first benchmark problem significant economic benefit was realized by finding the optimal setpoint signal after variance reduction. The process operation has been optimized at different confidence levels. Thanks to the efficiency of MADS the optimization has taken only 5 minutes; however, 100 iterations have been set in the Monte Carlo simulation. In case of the non-linear process the increase of economic throughput is not as significant as in the previous case, but with application of the proposed framework 5% profit increase can be obtained. The conditions of the optimization were similar to the previous case, 100 individual runs hve been set in the Monte Carlo simulation. Since the process model is non-linear the time consumption of the optimization is longer, almost one hour.

The application of the proposed methodology requires an existing process model with the description of the control system and detailed analysis of the process uncertainties. These modeling and analysis tools are widely available in advanced technologies thanks to the increasing interest for APC and Operator Training Systems (OTS). Another inevitable condition of application is the availability of an economic objective function. Although economic performance measures are frequently missing in current technologies, nowadays online economic performance monitoring is more and more indispensable. The aim of the application of these economic oriented monitoring tools is to avoid operations that are not economically or energetically optimal. Utilizing these performance assessment tools and integrating them with optimization the proposed framework is resulted.

ACKNOWLEDGMENTS

This work was supported by the European Union and financed by the European Social Fund in the frame of the TAMOP-4.2.1/B-09/1/KONV-2010-0003 and GOP-1.1.1-11-2011-0045 projects.

REFERENCES

1. K. H. Lee, E. C. Tamayo, and B. Huang, "Industrial implementation of controller performance analysis technology," Control Engineering Practice, vol. 18, no. 2, pp. 147–158, 2010.
2. J. Oakland, Statistical Process Control, Butterworth-Heineman, 2007.
3. M. Bauer and I. K. Craig, "Economic assessment of advanced process control—a survey and framework," Journal of Process Control, vol. 18, no. 1, pp. 2–18, 2008.
4. X. Chen, M. Heidarinejad, J. Liu, and P. D. Christofides, "Distributed economic mpc: application to a nonlinear chemical process network," Journal of Process Control, vol. 22, pp. 689–699, 2012.
5. K. H. Lee, B. Huang, and E. C. Tamayo, "Sensitivity analysis for selective constraint and variability tuning in performance assessment of industrial MPC," Control Engineering Practice, vol. 16, no. 10, pp. 1195–1215, 2008.
6. C. Zhao, Y. Zhao, H. Su, and B. Huang, "Economic performance assessment of advanced process control with LQG benchmarking," Journal of Process Control, vol. 19, no. 4, pp. 557–569, 2009.
7. C. E. García, D. M. Prett, and M. Morari, "Model predictive control: theory and practice—a survey," Automatica, vol. 25, no. 3, pp. 335–348, 1989.
8. R. Rubinstein and D. Kroese, Simulation and the Monte Carlo Method, Wiley-Interscience, 2008.
9. E. Borgonovo, M. Marseguerra, and E. Zio, "A Monte Carlo methodological approach to plant availability modeling with maintenance, aging and obsolescence," Reliability Engineering and System Safety, vol. 67, no. 1, pp. 61–73, 2000.
10. C. Audet and J. E. Dennis, "Mesh adaptive direct search algorithms for constrained optimization," SIAM Journal on Optimization, vol. 17, no. 1, pp. 188–217, 2006.
11. W. H. Ray, Advanced Process Control, McGraw-Hill, New York, NY, USA, 1981.
12. A. Prékopa, Stochastic Programming, Springer, 1995.

13. K. Marti, Y. Ermoliev, and G. Pflug, Dynamic Stochastic Optimization, Springer, 2004.
14. G. Kjellstrom, "Network optimization by random variation of component values," Ericsson Technics, vol. 25, no. 3, pp. 133–151, 1969.
15. M. A. Abramson, C. Audet, J. W. Chrissis, and J. G. Walston, "Mesh adaptive direct search algorithms for mixed variable optimization," Optimization Letters, vol. 3, no. 1, pp. 35–47, 2009.
16. V. Torczon, "On the convergence of pattern search algorithms," SIAM Journal on Optimization, vol. 7, no. 1, pp. 1–25, 1997.
17. N. L. Ricker, "The use of biased least-squares estimators for parameters in discrete-time pulse-response models," Industrial and Engineering Chemistry Research, vol. 27, no. 2, pp. 343–350, 1988.
18. A. Silva-Beard and A. Flores-Tlacuahuac, "Effect of process design/operation on the steady-state operability of a methyl methacrylate polymerization reactor," Industrial and Engineering Chemistry Research, vol. 38, no. 12, pp. 4790–4804, 1999.
19. B. R. Maner and F. J. Doyle, "Polymerization reactor control using autoregressive volterra-based mpc," AIChE Journal, vol. 43, no. 7, pp. 1763–1784, 1997.

Chapter 2

Sustainability Assessment of Chemical Processes: Evaluation of Three Synthesis Routes of DMC

Paula Saavalainen[1], Satish Kabra[2], Esa Turpeinen[1], Kati Oravisjärvi[1], Ganapati D. Yadav[2], Riitta L. Keiski[1], and Eva Pongrácz[3]

[1]Environmental and Chemical Engineering, Faculty of Technology, University of Oulu, Oulu, Finland
[2]Department of Chemical Engineering, Institute of Chemical Technology (ICT), Matunga, Mumbai 400019, India
[3]Thule Institute, NorTech Oulu, University of Oulu, Oulu, Finland

ABSTRACT

This paper suggested multicriteria based evaluation tool to assess the sustainability of three different reaction routes to dimethyl carbonate: direct synthesis from carbon dioxide and methanol, transesterification of methanol and propylene carbonate, and oxidative carbonylation of methanol. The first two routes are CO_2-based and in a research and development phase, whereas the last one is a commercial process.

The set of environmental, social, and economic indicators selected were renewability of feedstock, energy intensity, waste generation, CO_2 balance, yield, feedstock price, process costs, health and safety issues of feedstock, process conditions, and innovation potential. The performance in these indicators was evaluated with the normalized scores from 0 to +1; 0 for detrimental and 1 for favorable impacts. The assessment showed that the transesterification route had the best potential toward sustainability, although there is still much development needed to improve yield. Further, the assessment gave clear understanding of the main benefits of each reaction route, as well as the major challenges to sustainability, which can further aid in orienting development efforts to key issues that need improvement. Finally, it was concluded that a multicriteria analysis such as the one presented in this paper was a viable method to be used in the process design stage.

INTRODUCTION

In the last decades, sustainable development has become the cornerstone of environmental policy and a leading principle for resource management. The widely used definition of sustainable development is that of the United Nations' Brundtland Commission [1]: "Development that meets the needs of the present without compromising the ability of future generations to meet their own needs." In corporate terms, sustainability can be summarized as the "triple bottom line" (TBL) success [2], which implies that firms have to maintain and grow their economic, social, and environmental capital base, while actively contributing to sustainability in the political domain [3, 4].

One of the key challenges of sustainable development is that it demands new and innovative choices and ways of thinking. Innovations in technology are challenging organizations to make new choices in their operations, products, and activities that impact the earth and people as well as economics [5]. There is, however, no standard method for measuring the triple bottom line success of technological innovations at the design phase and the principles to achieve sustainability by themselves are insufficient to create the right framework for design towards sustainability [6]. It would be useful to have a screening tool to assess how a new product or process under development would perform in terms of sustainability, or compared with a commercial process. Although there are various international efforts to measure sustainability, only a few of them have an integral approach taking into account environmental, economic, and social aspects. In most cases, the focus is on one of the three aspects [7]. For example, Life Cycle Assessment (LCA) is used to evaluate the environmental performance of products, but it concentrates on environmental impacts only [8]. As well, environmental impact assessment (EIA), a procedural tool for the design phase, only evaluates the environmental implications of decisions [9].

In order to fully evaluate the sustainability of new process routes, there is a need for a comprehensive evaluation of the environmental, economic, and social impacts of these new routes at an early process design stage. The paper suggests using multicriteria assessment for sustainability assessment and demonstrates its use in assessing a novel carbon dioxide-based reaction route to dimethyl carbonate (DMC).

SUSTAINABILITY ASSESSMENT METHODOLOGIES

There are a number of sustainability assessment methodologies evaluating the performance of industrial facilities. The World Business Council for Sustainable Development [10], the Global Reporting Initiative [5], and development of standards [11] are key drivers for adopting sustainability management in industries.

The most extensive work in terms of sustainability assessment has been done by the Global Reporting Initiative (GRI). GRI is a nongovernmental organization that aims at driving sustainability and has developed an environmental, social, and governance (ESG) reporting framework to be used worldwide. GRI version 4 on Sustainability Reporting Guidelines defines the principles and indicators that organizations can use to measure and report their economic, environmental, and social performance. Many companies use these indicators while publishing their annual or environmental reports. GRI is committed to continuously improve and increase the use of the guidelines which are available to the public [5].

The American Institute of Chemical Engineers (AIChE) has defined the AIChE Sustainability Index (SI) to measure the sustainability performance of representative companies in chemical industry [12]. The AIChE SI uses publicly available data on the companies' strategic commitment, sustainability innovation, environmental performance, safety performance, product stewardship, social responsibility, and value chain management to measure their sustainability performance. Metrics to measure the "greenness" of the companies' chemistry have been developed by the American Institute of Chemical Engineers' Center for Waste Reduction Technologies (AIChE/CWRT) assessing material intensity, energy intensity, water consumption, toxic release, and pollutant effects. The metrics developed are simple, understandable, easy to reproduce, and comparable [13]. They take into notice also the social aspects of sustainability by considering the health effects the chemicals used/produced have. However, they are developed for companies and are adjusted for existing process improvements rather than for a new process design.

Similarly, the Institution of Chemical Engineers (IChemE) has developed a set of metrics to enable process industry companies to measure and report progress along the path of sustainable development [14]. The Sustainable Development Progress Metrics are intended to help companies to set targets and develop internal standards and to monitor their progress in time [15]. The IChemE metrics are divided into environmental, economic, and social indicators. The environmental indicators are concentrating on resource use by considering how much energy, material, and water are consumed and land is used. Also atmospheric, aquatic impacts, and impacts on land caused by emissions, effluents, and waste are taken into notice. The economic indicators are concentrating on the profit gained, value added and taxes paid, and investments made by the company. The social indicators are considering the employment situation, health and safety at work, and also impacts to society. Not all the metrics proposed are valid in every case and it is up to the companies to decide which of the metrics are relevant for them. Key indicators have to be chosen from each of the aspects of sustainability to give a balanced view of the sustainability performance [15]. Whilst the IChemE metrics account for all three aspects of sustainability, they are meant as a sustainability management tool for companies, aiming at enhancing their sustainability performance, and are not suitable for assessing processes under development.

In terms of sustainability guidance for chemicals and chemical process design, Green Chemistry was developed to reduce or eliminate negative environmental impacts [16, 17]. The 12 Principles of Green Chemistry have been a cornerstone of environmentally conscious chemical process design since the late 1990s. Green Chemistry had been suggested to be used as a pollution prevention tool as it applies innovative scientific solutions to real-world environmental situations [18]. However, the assessment range of Green Chemistry does not cover the full depth of sustainability. As it was its original purpose, its emphasis is on reducing the toxicity of chemical products and driving inherently safer chemistry.

Protection of human health and the environment from chemicals and associated risks is also the goal of the European REACH (The Registration, Evaluation, Authorization and Restriction of Chemicals) regulation, which came into force in 2007. It renewed and upgraded the previous chemicals regulatory framework of the European Union (EU) [19], in order to ensure that there is free circulation of substances on the internal market and to enhance competitiveness and innovation. REACH confirms that industries are responsible for both assessing and managing the risks associated with chemicals, giving suitable safety information of chemicals to users, and promoting alternative testing methods [20]. About 143 000 chemicals marketed in the EU were preregistered by the December 1, 2008, deadline in REACH. The registration document of chemicals under REACH includes general information, safety data sheets (SDS), chemical safety report (CSR), and chemical safety assessment (CSA). Testing for health hazards under REACH includes acute toxicity, skin corrosion and irritation, serious eye damage and irritation, skin or respiratory sensitizer effect, mutagenic or carcinogenic impacts, toxicity for reproduction, specific target organ toxin in single exposure, specific target organ toxin in repeated exposure, and aspiration hazard [21, 22].

Table 1 summarizes some of the main evaluation guidelines or indicators used in the mentioned assessment processes. All methods outlined in Table 1 take into notice some key aspects of sustainability and clearly intend to evaluate triple bottom line success. Indicators that are possible to assess in the design phase and would give a good signal of sustainability performance are bolded. In terms of sustainability assessment of chemical processes in the design phase, Green Chemistry is the most thorough; however, should it be used to assess sustainability, it is recommended to extend it with social and economic indicators of GRI, AIChE, and IChemE to give a comprehensive measure of sustainability.

Table 1: Main principles/evaluation guidelines of reviewed assessment methods [5,12–16]

	GRI	AIChE	IChemE	Green Chemistry
Environmental performance	Materials	(i) Resource use	Resource usage	Prevent waste
	Energy	(a) Energy	(i) Energy	Use renewable feedstock
	Water	(b) Materials	(ii) Material	
	Biodiversity	(c) Renewables	(iii) Water	Avoid chemical derivatives
	Emissions	(d) Water	(iv) Land	Catalysts
	Effluents and waste	GHG emissions	Emissions, effluents, and waste	Product degradability
	Products and services	Waste, wastewater		
	Compliance	Compliance management		
	Transport	Value chain management		
	Suppliers			
Economic performance	Economic performance	(i) Sustainability innovation	Profit, value, tax	Maximise atom economy
	Market presence	(ii) Strategic commitment to sustainability	(i) Investments	Increase energy efficiency
	Procurement practices			
Social performance	Labor practices	Social responsibility	Workplace	Less hazardous chemical syntheses
	(i) Employment	(i) Stakeholder partnership	(i) Employment	Safer chemicals, products, solvents, and reactions
	(ii) Health and safety	(ii) Social investment	(ii) Health and safety	
	(iii) Innovation and knowledge potential	(iii) Image in the community	society	Accident prevention and real time analysis
	(iv) Diversity and equality	Product stewardship		
	society	(i) assurance system		
	(i) Acceptability and social dialogue	(ii) risk communication		
	Human rights	(iii) legal proceedings		

Following the recommendation of IChemE, we selected key indicators from each of the aspects of sustainability to give a balanced

view of the sustainability performance. The suggested indicators are as follows:
- environmental indicators:
  feedstock renewability,
  energy intensity,
  waste generation,
  CO_2 balance,
- economic indicators:
  yield,
  feedstock price,
  process costs,
- social indicators:
  process conditions,
  chemicals safety,
  innovation potential.

These indicators were selected as they can be assessed based on reaction routes as well as laboratory scale experiments and thermodynamic simulations. We propose that these 10 indicators are a necessary and sufficient set of meters for screening purposes at the design phase and give a balanced view of chemical process sustainability.

ASSESSMENT OF DMC PRODUCTION ROUTES

Dimethyl carbonate (DMC, $(CH_3)_2CO$) is an important chemical intermediate that can be used as a fuel additive and a polar solvent in the chemical industry. The production of DMC has received increasing attention over the least years [23–28]. There are several methods for the synthesis of DMC, such as phosgenation of methanol, oxidative carbonylation of methanol, transesterification method, and esterification of carbon dioxide with methanol [29–31]. In this paper, three reaction routes for DMC synthesis are evaluated. The reaction routes are outlined in Table 2.

Table 2: The assessed reaction routes for DMC production

Route A: direct synthesis from carbon dioxide and methanol
$CO_2 + 2CH_3OH \rightarrow (CH_3O)_2CO + H_2O$
Route B: transesterification of methanol and propylene carbonate using ionic liquid (IL) as a catalyst
$C_3H_6O + CO_2 \rightarrow C_4H_6O_3$
$C_4H_6O_3 + 2CH_3OH \rightarrow (CH_3O)_2CO + C_3H_8O_2$
Route C: oxidative carbonylation of methanol (ENiChem)
$2CH_3OH + 1/2O_2 + CO \rightarrow (CH_3O)_2CO + H_2O$

All three routes provide a safer alternative for the primary synthesis pathway, the "phosgene route" $COCl_2 + 2CH_3OH$ $(CH_3O)_2CO + 2HCl$. The use of phosgene route is phased out from the commercial processes, as phosgene is one of the most acutely toxic substances used in industrial scale. As this route presents inherent hazards and potential environmental problems in handling and waste disposal [23], it is crucial that it is replaced by a more sustainable method.

Route A is currently in academic research phase. This route is particularly attractive for being CO_2-based. Generally speaking, carbon dioxide (CO_2) can be considered as an environmentally friendly and widely available feedstock, available as a waste emission of industrial processes. Chemical utilization of CO_2 for DMC manufacture would be a means to turn this waste into a nonwaste, allowing us to view CO_2 as a useful resource. It has been reported earlier that CO_2-based synthesis processes are meeting many of the provisions for environmental, economic, and social sustainability [32]. Therefore, much academic research has concentrated on the search for benign by design synthesis involving CO_2 as a raw material [33]. The synthesis of carbonic esters is one example [34]. The expectation is that the CO_2-based DMC production routes have significant potentials toward sustainable production. However, there are also numerous challenges of CO_2 utilization [32]; therefore, long-term research efforts for acquiring the necessary knowledge in its chemical reactivity are needed.

Route B is also an attractive "carbon-friendly" route, due to using CO_2 as a reaction feedstock. However, the complexity of a two-step process, the use of toxic propylene oxide, and the coproduction of propylene glycol make this process demanding. The challenge in both Routes A and B is that scale-up of the production would not be economically feasible at the moment.

The commercial route (ENiChem), Route C, is based on the catalytic oxidative carbonylation of methanol. It offers operational and environmental advantages, for example, fewer side products, better atom economy, and safer production comparing to the phosgene route [23, 33] but it is not responding to the current demand of DMC.

In terms of "measuring" innovation potential, we performed a literature review using a simple keyword search in Science Direct to evaluate the volume of publications and calculated the percentage of recent publications (2012 or later) of the 50 most relevant publications. Our reasoning is that the volume of publications is indicative of the level of knowledge potential, and the high percentage of recent publications indicates intensified academic interest, which will contribute to the renewal of science and is more likely to drive innovation.

PROCESS SIMULATIONS

Mass and energy balances of process routes were calculated by Aspen Plus simulations. Process flow sheets are presented in Figures 1 and 2. In order to make comparison of processes simple and appropriate the flow sheets were designed as similar as possible. Processes A and C consist of a reactor, gas separation unit, flash separator, and two columns. Process B is composed of two reactors and three columns. The reactors used were modeled as stoichiometric reactors based on known fractional conversion of a certain component. Radfrac model was used in separation units. The routes were assumed to be ideal (no mass and heat losses and no pressure drops and ideal component properties). The foundation for the calculation was stoichiometric, based on reaction equation, 1 kmol of each. The process conditions for the inlet stream were as follows: temperature 20°C and pressure 1 bar. The outlet stream temperature was set at 20°C and pressure at 1 bar. Concentration of DMC after purification was adjusted to 85 vol-%.

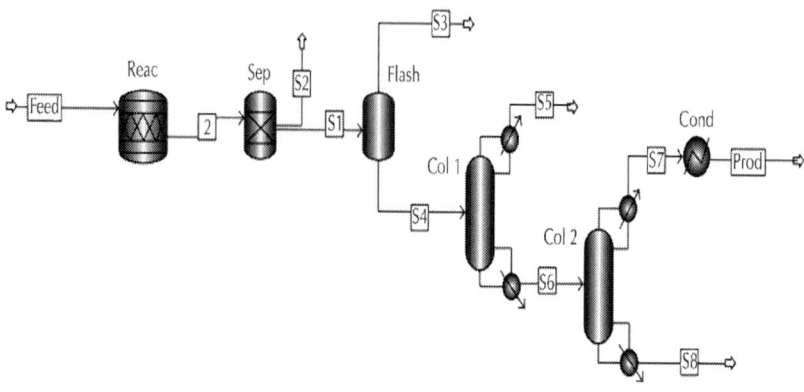

Figure 1: Process flow sheet for Routes A and C.

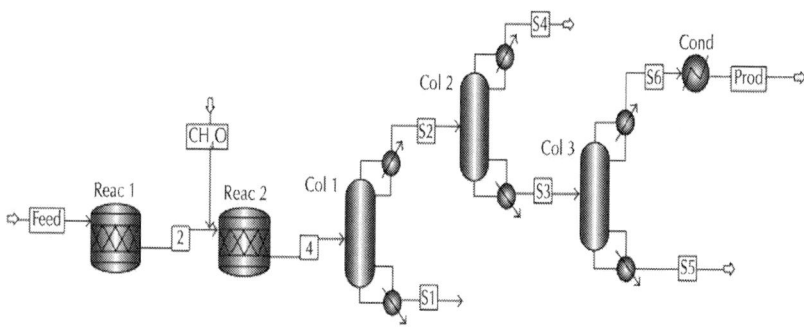

Figure 2: Process flow sheet for Route B.

Detailed descriptions of the process units and conditions are presented in Tables 3–5.

Table 3: Process description for Route A

Process unit	Type	Conditions	Notes
Reac	Stoichiometric reactor	T = 50°C, P = 150 bar	Conversion of CH_3OH = 8.3%
Sep	Component separator	Split fraction of CO_2 = 100%	Separation of unreacted CO_2
Flash	Flash separator	T = 97°C, P = 3 bar	

Col 1	RadFrac column	15 stages, distillate rate = 1.421, reflux ratio = 5	Separation of methanol
Col 2	RadFrac column	15 stages, distillate rate = 0.07, reflux ratio = 9	Concentration of DMC after distillation = 85.5%
Cond	Cooler	T = 20°C, P = 1 bar	Cooling of DMC

Table 4: Process description for Route B

Process unit	Type	Conditions	Notes
Reac 1	Stoichiometric reactor	T = 100°C, P = 140 bar	Conversion of propylene oxide = 100%
Reac 2	Stoichiometric reactor	T = 150°C, P = 1 bar	Conversion of methanol = 5.25%
Col 1	RadFrac column	15 stages, distillate rate = 2, reflux ratio = 5	Separation of propylene carbonate
Col 2	RadFrac column	15 stages, distillate rate = 1.895, reflux ratio = 5	Separation of methanol
Col 3	RadFrac column	15 stages, distillate rate = 0.06, reflux ratio = 5	Concentration of DMC after distillation = 85.9%
Cond	Cooler	T = 20°C, P = 1 bar	Cooling of DMC

Table 5: Process description for Route C

Process unit	Type	Conditions	Notes
Reac	Stoichiometric reactor	T = 120°C, P = 27 bar	Conversion of CH_3OH = 16.49%
Sep	Component separator	Split fraction of CO and O_2 = 100%	Separation of unreacted CO and O_2
Flash	Flash separator	T = 99.1°C, P = 3 bar	
Col 1	RadFrac column	15 stages, distillate rate = 1.307, reflux ratio = 5	Separation of methanol
Col 2	RadFrac column	15 stages, distillate rate = 0.148, reflux ratio = 10	Concentration of DMC after distillation = 85.4%
Cond	Cooler	T = 20°C, P = 1 bar	Cooling of DMC

ASSESSMENT PROCESS FOR REACTION ROUTES

The assumptions for all 3 reaction routes used in the assessment are summarized in Table 6 and the simulation results are gathered in Table 7. Process details (reactants, products, solvents, wastes, catalyst, temperature, pressure, conversion, and selectivity) were taken from the articles or/and academic theses [25, 35–39]. In addition, in Table 6 also the data for the literature review was included.

Table 6: Facts and assumptions regarding the three reaction routes

	Route A: direct synthesis from carbon dioxide and methanol	Route B: transesterification of methanol and propylene carbonate	Route C: oxidative carbonylation of methanol (ENiChem)
Reaction route (stoichiometric feed [kmol])	$CO_2 + 2CH_3OH \rightarrow (CH_3O)_2CO + H_2O$	$C_3H_6O + CO_2 \rightarrow C_4H_6O_3$ $C_4H_6O_3 + 2CH_3OH \rightarrow (CH_3O)_2CO + C_3H_8O_2$	$2CH_3OH + 1/2 O_2 + CO \rightarrow (CH_3O)_2CO + H_2O$
Atom economy [%] (theoretical)	83.3	60.8	83.3
Raw materials	CO_2 and CH_3OH	CH_3OH, CO_2, $C_3H_6O_3$ Intermediate: $C_4H_6O_3$	CH_3OH, O_2 and CO
Supply chain	CH_3OH from natural gas CO_2 separated from flue gas by absorption (MEA)	CH_3OH from syngas CO_2 separated from flue gas by absorption (MEA) Propylene oxide from H_2O_2 and propene	CH_3OH from natural gas O_2 from air (distillation) CO from natural gas
Solvents and auxiliary chemicals	IL101 as a promoter	No solvents or auxiliary chemicals	No solvents or auxiliary chemicals
Catalyst	CHT-HMS	First step: ion exchange resin D201 Second step: IL 103	$CuCl_2$
By-products and coproducts	H_2O unreacted CH_3OH	Propylene glycol, unreacted CH_3OH, and propylene carbonate	H_2O, unreacted CH_3OH, O_2, CO, and H_2O
Waste and emissions	Methylformate, unreacted CO_2	No wastes	No wastes

Process conditions	Pressure 150 bar Temp. 50°C Supercritical CO_2	First step:	Pressure 140 bar Temp. 100°C Supercritical CO_2	Pressure 27 bar Temp. 120°C
		Second step:	Pressure 1.01325 bar Temp. 150°C	
Health and safety issues	Methyl formate (i) is extremely flammable (ii) is harmful if swallowed or inhaled (iii) causes serious eye irritation (iv) may cause respiratory irritation	Propylene oxide (i) is extremely flammable (ii) is harmful if swallowed, inhaled, or came in contact with skin (iii) may cause respiratory irritation (iv) may cause genetic defects and cancer		CO is (i) flammable (ii) toxic for human CO and O_2 must be fed at a carefully controlled rate to avoid the risk of explosion
Volume of articles	542	129		76
Percentage of recent publications	25%	50%		8%

Table 7: Simulation results and cost calculations of reaction routes

	Route A	Route B	Route C
Conversion of MeOH [%]	9.16	(1) Step: 100 (2) Step: 10.5	17
Selectivity to DMC [%]	90.56	(1) Step 100 (2) Step 50	97
Yield [%]	5.99	5.15	12.64
Atom economy [%] (Real)	7.19	8.47	15.17
Amount of DMC (kmol)	0.06	0.051	0.126
Concentration of DMC (vol-%)	85.6	85.9	85.4
CO_2 emissions [kmol]	0.92	0	—
CO_2 consumption [kmol]	0.08	1.00	—
CO_2 balance	0.84	1	0
Energy consumption (specific) [MJ/DMC produced]	1152.8	−746.3	131.5

Energy consumption (Aspen) [MJ]	69.17	−38.06	16.57
Costs of feedstock (/kmol)	CO_2: 24.8	CH_3OH: 28.9	CH_3OH: 28.9
	CH_3OH: 28.9	CO_2: 24.8	O_2: 13.4
		C_3H_6O: 72.6	CO: 199
	tot. 53.7	tot. 126.3	tot. 241.3
Operational costs (feedstock + process) [ ]	53.7 + 18.7 = 72.4	126.3 − 10.3 = 116	241.3 + 4.5 = 245.8
Operational costs (feedstock + process)/DMC produced / kmol	1206.7	2274.5	1950.8
Treatment cost/waste disposal cost	High disposal cost of methylformate Water can be discharged to drain	By-product can be sold	Water can be discharged to drain

In reaction Route A, fossil fuel based raw materials are used, where it is assumed that methanol is produced with carbon monoxide (CO), CO_2, and hydrogen (H_2). Reaction Route B uses oil refinery products and CO_2 as raw material. In reaction Routes A and B, commercial catalytic materials are under development. The academic research toward reaction Route A uses calcined hydrotalcite on hexagonal mesoporous silica (CHT-HMS) as a catalyst with an IL promoter. In reaction Route B, ion exchange resin and ionic liquid (IL) are used as catalyst material. Finally, in Route C, commercial copper chloride catalysts are used. Catalytic materials for routes under development should be chosen for the assessment in order to minimize the environmental impact of catalyst materials, that is, enhancement of reaction activity and selectivity and stability of the catalyst, as well as environmentally benign catalytic materials.

In reaction Routes A and B, optimal reaction conditions such as temperature and pressure are not yet resolved as these routes are still under development; however, they are expected to be rather high and supercritical CO_2 is used. Reaction conditions should further be developed so that temperature and pressure are optimized at a lower level to minimize risks and environmental impacts. Reactions in Route C are using lower pressure, but higher temperature. The environmental benefit of this is to be highlighted, when compared with routes under research.

In reaction Routes A and C, only water is produced as a by-product. In Route A also small amounts of methylformate are produced. In Route

B, toxic propylene glycol is produced. Propylene glycol is valuable from a commercial point of view; however its possible utilization needs to be considered at the design phase.

Atom economy is the best for the methanol-based reaction Routes A and C. However, it needs to be assessed if the atom economy benefit overweighs other impacts of the reactions. Reaction route C uses CO as a raw material, the production of which is rather energy demanding. In reaction Routes A and B, the yield is very low because of low conversion of methanol. This highlights the need for research for more efficient catalyst materials.

Prices of feedstock were acquired from chemical suppliers. Total operating costs were calculated by summing prices of feedstock and energy consumption of the process. Energy consumption of the process (MJ) was converted to euros by rate of 0.27 e/MJ (Eurostat). Capital costs were left out of considerations because all the cases are quite similar, and thus they were assumed to be equal in capital costs.

SUSTAINABILITY ASSESSMENT OF DMC ROUTES

Sustainability assessments are multicriteria based evaluations, which necessitate the inclusion of a wide variety of data typology with various certainty degrees. In this paper, we use multicriteria assessment (MCA) to perform the evaluation of the three DMC routes. Various multicriteria decision analysis methods have been put forward as an excellent candidate to perform sustainability assessment recently, and a variety of applications have emerged [40]. MCA is formal approach that takes into account multiple criteria in order to help making decisions that matter [41]. MCA stands in contrast to single goal optimization and approaches which, when using "unifying units," may offset poor performance of one criterion by good performances of another criterion, therefore allowing for substitution and compensability between criteria [42].

MCA methods require data to be normalized in order to obtain comparable scales. A common method is the ratio normalization that attributes value 1 to the best performance on a criterion and a proportional value to the other performances [43]. The objective of this method is to provide an easy to use screening tool for assessment and comparison in the design phase, in order to point out key aspects that need to be improved on or further explored. In some cases, we have amended this method in a way that the most preferred performance was valued 1 while detrimental performance was valued 0 and, if applicable, the third value normalized in between. In some cases we were reduced to qualitative evaluation, assigning 1 for best, 0 for worst, and 0.5 for medium values.

Table 8 lists the normalized values of selected indicators. Routes B and C use one-third of raw materials from oil refinery products, and 50% of raw materials in Route A are renewable. Values are normalized accordingly. In terms of energy demand, Route B releases 746.3 MJ energy, while reaction Routes A and C consume energy. We assigned Route A (1152.2 MJ/DMC production) a 0 value, to route B the value 1, and normalized the consumption of Route C (131.5 2 MJ/DMC) to the value 0.54. Only Routes A and B are CO_2-based. Route C has therefore no direct CO_2 implication. Route B consumes CO_2, while Route A generates it. We assigned Route B the value of 1 and Route C value 0 and normalized Route A in between. In case of wastes, Routes B and C produce no wastes, while Route A produces low amounts of methyl formate. Therefore, A is valued 0 while B and C are valued 1.

Table 8: Sustainability indicator values

	Route A	Route B	Route C
Environmental indicators			
Feedstock renewability	1	0.67	0.67
Energy intensity	0	1	0.54
CO_2 balance	0.84	1	0
Wastes	0	1	1
Economic indicators			
Yield	0.48	0.41	1
Feedstock price	1	0.67	0

Process costs	1	0.3	0
Social indicators			
Process conditions	0	0	0.06
Chemicals safety	1	0.5	0
Innovation potential	0.8	1	0.1
	6.12	6.55	3.37

The yield in reaction Route C is the highest (1), as expected from a commercial process. The yields of Routes A and B are moderate; normalized values are 0.48 and 0.41. Both have the potential to enhance the selectivity and yield as well. The yield of DMC in the process Route A can be improved by circumventing the thermodynamic limitations. The water generated in the process can be chemically trapped as discussed by Eta et al. [44] and thus the equilibrium can be shifted in the forward direction for a higher yield of DMC. The feedstock costs of raw materials for Route C are highest, and therefore it is valued 0; for Route A the costs are the lowest, thus valued 1, and route B has a normalized value of 0.67. The real processing costs are difficult to assess for Routes A and B, which are in the design phase; therefore, theoretical figures of operational costs were used. The assessment was based on the composite costs of feedstock and energy, divided by the amount of produced DMC. Based on this, Route C is the most expensive (0), Route A is the cheapest (1), and Route B is moderate and has normalized value 0.3.

For process conditions, the process temperature and pressure were evaluated. Room temperature (21°C) and atmospheric pressure (1 bar) were considered the safest, which would be valued 1. We assigned 0 for highest temperature 150°C (Route B) and the highest pressure 150 bar (Route A) normalized the other values, 50°C in Route A 0.61 and 120°C in Route C 0.07 and 140 bar in Route B 0,07 and 27 bar in Route C 0,92. These values were multiplied for a composite value. Health and safety issues are most severe for reaction Route C (0) due to the use of CO, less severe for Route B 0.5 that is using organic solvents, and benign in the case of Route A (1). Innovation potential was valuated based on the volume of articles published on these production methods and the percentage of recent papers. Most articles were written on subject related to Route A but only 20% of the relevant were recent, indicating a receding interest. In case of Route B, the volume of publications is moderate, but 50% of the most relevant are recent, which indicates this is of rising scientific relevance. The innovation potential of Route B was evaluated highest (value 1) and Route A was normalized to 0.8, while Route C with the fewest volume and least recent publications was valued 0.1.

The results of this comparative assessment are presented in Figure 3. Route B (red line) seems to be the most positive from environmental and social points of view; the only negative issue is the relatively highest safety risk in terms of process conditions, but it performs best in terms of low energy consumption and CO_2 balance as it consumes CO_2. Route A (blue line) seems to have some potential toward economic and social sustainability; however, in terms of environmental sustainability, it has some shortcomings, such as high energy consumption and waste generation. The commercial process (green line) performs best in terms of yield, which is expected from a mature process; however, it has the worst social sustainability performance and it is also based on nonrenewable feedstock. Table 9 summarizes the benefits and challenges of the three routes.

Table 9: Summary of sustainability assessment, benefits, and challenges to sustainability

	Benefits driving sustainability	Challenges to sustainability
Route A	(i) CO_2 used as feedstock (ii) High volume of academic papers (iii) Lowest total cost per produced DMC (iv) Safest chemicals	(i) Low yield, low conversion (ii) Methylformate as waste (iii) High energy consumption (iv) High pressure process
Route B	(i) CO_2 is a feedstock in the first step (ii) Valuable by-product (iii) Energy win (iv) Intensified academic research (v) Consuming all CO_2	(i) Oil refinery product used as feedstock (ii) Low yield (iii) Propylene oxide use is an inherent risk (iv) Highest total cost per produced DMC (v) High process temperature and pressure

Sustainability Assessment of Chemical Processes: Evaluation of Three... 45

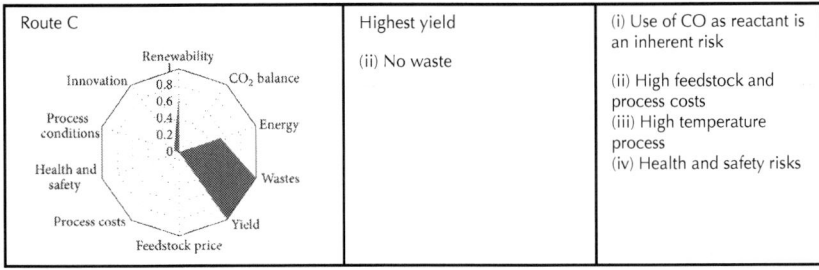

Route C	Highest yield	(i) Use of CO as reactant is an inherent risk
	(ii) No waste	(ii) High feedstock and process costs
		(iii) High temperature process
		(iv) Health and safety risks

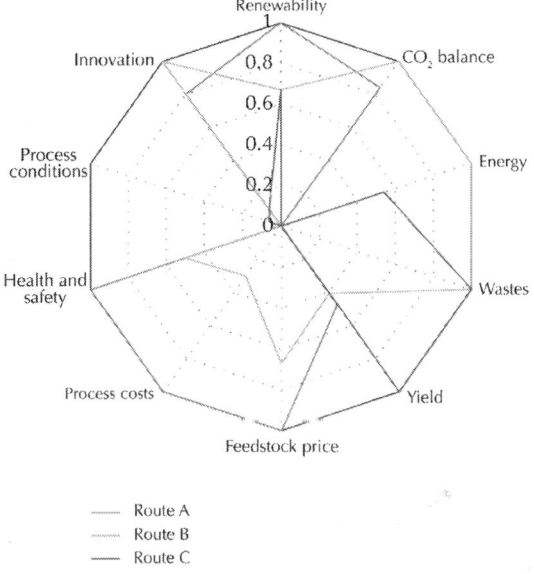

------ Route A
------ Route B
―― Route C

Figure 3: Comparison of the three DMC reaction routes.

In summation, it can be asserted that Route B has the best potential toward sustainability, although there is still much research needed to improve yield and conversion and thus reduce the amounts of wastes. In this case, use of a better catalyst would be further useful and add to sustainability positively. As its shape also indicates, Route A is very conflicting, as it has almost equal amounts of positive and negative factors. Many of the challenges are, however, difficult to overcome, such as the use of nonrenewable feedstock and yield stemming from low theoretical atom economy. In the commercial process (Route C), the toxicity of the reactant and the high feedstock price and production cost are failings that may not be further improved.

CONCLUSIONS

In order to drive sustainability in the chemical industry, there is a need for a methodology capable of assessing the impact of new choices in products, processes, and operations at the design phase. Most sustainability assessment methods are meant to be tools of sustainability management on the corporate level. There are tools available to assess the environmental performance of products, such as Life Cycle Assessment; however they do not take into account economic and social implications. For the assessment tool presented in this paper, the principles of Green Chemistry were used as the basis of evaluation. The objective of Green Chemistry to promote safer chemistry is its strength in terms of driving sustainability, but it also has some limitations. It was meant to provide guidelines for design rather than being an assessment or a screening tool. Sustainability assessments are multicriteria based evaluations; therefore, we used multicriteria assessment (MCA) to perform the evaluation of the three DMC routes. Cross-referencing the Green Chemistry principles with established sustainability assessment and reporting methods (Global Reporting Initiative, AIChE, IChemE, and REACH), this paper suggested a manageable list of factors considered necessary and sufficient to gain an overview of impacts toward sustainability.

The renewable nature of feedstock, energy intensity, CO_2 balance, and waste generation were evaluated as environmental indicators. To assess economic performance, yield, price of feedstock, and process and production costs were selected. In terms of social sustainability, process conditions and chemicals safety were assessed, the latter using the guidelines of REACH. In addition, innovation and knowledge potential was assessed based on the volume and novelty of scientific publications recently published. It was argued that these factors could be assessed based on reaction routes, laboratory scale experiments and results, as well as thermodynamic simulations. As MCA methods require data normalization, we used the common method of 0-1 attribute values, 1 being the best and 0 the worst.

Of the three reaction routes to DMC, two are CO_2-based still in a research phase. The assessment indicated that transesterification has the best potential toward sustainability, although there is still much research needed to improve yield and selectivity. Direct synthesis from CO_2 and methanol has many positive attributes, but an almost equal amount of negative factors. The commercial process, oxidative carbonylation, has performed worst in terms of sustainability, the toxicity of the reactant, and the high feedstock cost providing the major limitations to further improvement.

It can be concluded that the assessment allowed pointing out the main benefits of each reaction route, as well the major challenges to sustainability. This can further aid in orienting development efforts to key issues that need to be improved. Further, it can be asserted that, of the established sustainability tools our method builds on, Green Chemistry holds the most potential for chemical industry research and development. Green Chemistry is well known and trusted amongst chemical engineers and has practical tools and guidelines developed for process designers. Finally, it is suggested that multicriteria assessment can be used as a sustainability assessment method in the process design stage.

ACKNOWLEDGMENTS

This work was performed within the collaborative project "Sustainable Catalytic Syntheses of Chemicals using Carbon Dioxide as Feedstock (GreenCatCO2)" supported by Department of Science and Technology, Government of India (DST-GOI), and The Academy of Finland. The authors would like to thank the Academy of Finland (Project nos. 129173 (SUSE) and 140122 (GreenCatCO2)) and the Finnish Funding Agency for Technology and Innovation, Tekes (Project no. 40313/09 (Fermet)) for financial support. Ganapati D. Yadav also thanks DST for J.C. Bose National Fellowship and received support from R.T. Mody Distinguished Professor Endowment.

REFERENCES

1. WCED (World Commission on Environment and Development), "United Nations General Assembly document A/42/427," in Our Common Future, Oxford University Press, Oxford, UK, 1987.
2. J. Elkington, Cannibals with Forks: The Triple Bottom Line of 21st Century Business, New Society Publishers, Stoney Creek, Conn, USA, 1998.

3. T. Dyllick and K. Hockerts, "Beyond the business case for corporate sustainability," Business Strategy and the Environment, vol. 11, no. 2, pp. 130–141, 2002. · ·
4. J. B. Bowell, "Sustainability metrics, indicators, and indices for the process industries," in Sustainable Development in the Process Industries: Cases and Impact, J. Harmsen, Ed., pp. 5–23, John Wiley & Sons, 2010.
5. The GRI Sustainability Reporting Guidelines, 2013, https://www.globalreporting.org/resourcelibrary/GRIG4-Part1-Reporting-Principles-and-Standard-Disclosures.pdf.
6. J. García-Serna, L. Pérez-Barrigón, and M. J. Cocero, "New trends for design towards sustainability in chemical engineering: green engineering," Chemical Engineering Journal, vol. 133, no. 1–3, pp. 7–30, 2007. · ·
7. R. K. Singh, H. R. Murty, S. K. Gupta, and A. K. Dikshit, "An overview of sustainability assessment methodologies," Ecological Indicators, vol. 15, no. 1, pp. 281–299, 2012. · ·
8. M. Aresta and M. Galatola, "Life cycle analysis applied to the assessment of the environmental impact of alternative synthetic processes. The dimethylcarbonate case: part 1," Journal of Cleaner Production, vol. 7, no. 3, pp. 181–193, 1999. · ·
9. Directive 2011/92/EU of the European parliament and of the council of 13 December 2011 on the assessment of the effects of certain public and private projects on the environment, http://eur-lex.europa.eu/LexUriServ/LexUriServ.do?uri=OJ:L:2012:026:0001:01:EN:HTML.
10. World Business Council for Sustainable Development (WBCSD), Signals of Change: Business Progress Toward Sustainable Development, World Business Council for Sustainable Development (WBCSD), Geneva, Switzerland, 1997.
11. OECD—Organisation for Economic Co-operation and Development, "An Update of the OECD Composite Leading Indicators Short-Term Economic Statistics Division, Statistics Directorate/OECD," 2002, http://www.oecd.org.
12. The (AIChE) Sustainability Index: The Factors in Detail, 2009, http://www.aiche.org/resources/publications/cep/2009/january/aiche-sustainability-index-factors-detail.

13. D. Tanzil, G. Ma, and B. R. Beloff, "Automating the sustainability metrics approach," in Proceedings of the AIChE Spring Meeting, New Orleans, La, USA, April 2004.
14. M. Wilkinson, "Sustainable development and IChemE," Process Safety and Environmental Protection, vol. 78, no. 4, p. 236, 2000. ·
15. IChemE sustainability metrics, sustainable development progress metrics recommended for use in the process industries, 2001,http://nbis.org/nbisresources/metrics/triple_bottom_line_indicators_process_industries.pdf.
16. P. T. Anastas and J. C. Warner, Green Chemistry: Theory and Practice, Oxford University Press, New York, NY, USA, 1998.
17. P. T. Anastas and M. M. Kirchhoff, "Origins, current status, and future challenges of green chemistry,"Accounts of Chemical Research, vol. 35, no. 9, pp. 686–694, 2002. · ·
18. J. B. Manley, P. T. Anastas, and B. W. Cue Jr., "Frontiers in Green Chemistry: meeting the grand challenges for sustainability in R&D and manufacturing," Journal of Cleaner Production, vol. 16, no. 6, pp. 743–750, 2008. · ·
19. W. Lilienblum, W. Dekant, H. Foth et al., "Alternative methods to safety studies in experimental animals: role in the risk assessment of chemicals under the new European Chemicals Legislation (REACH)," Archives of Toxicology, vol. 82, no. 4, pp. 211–236, 2008. · ·
20. Regulation (EC) No 1272/2008 of the European parliament and of the council of 16 December 2008 on classification, labelling and packaging of substances and mixtures, amending and repealing Directives 67/548/EEC and 1999/45/EC, and amending Regulation (EC) No 1907/2006, http://eur-lex.europa.eu/LexUriServ/LexUriServ.do?uri=OJ:L:2008:353:0001:01:EN:HTML.
21. Regulation of the European parliament and of the council on classification, labelling and packaging of substances and mixtures, amending and repealing directives 67/548/EEC and 1999/45/EC, and amending regulation (EC) No 1907/2006, http://echa.europa.eu/fi/addressing-chemicals-of-concern/harmonised-classification-and-labelling/annex-vi-to-clp.

22. C&L Inventory database, 2014, http://echa.europa.eu/fi/information-on-chemicals/cl-inventory-database;jsessionid=F950E8759BC3897F24C962972266BCD9.live1.
23. D. Delledonne, F. Rivetti, and U. Romano, "Developments in the production and application of dimethylcarbonate," Applied Catalysis A: General, vol. 221, no. 1-2, pp. 241–251, 2001. · ·
24. H. M. Wang, H. Wang, N. Zhao, W. Wei, and Y. Sun, "High-yield synthesis of dimethyl carbonate from urea and methanol using a catalytic distillation process," Industrial and Engineering Chemistry Research, vol. 46, no. 9, pp. 2683–2687, 2007. · ·
25. M. A. Pacheco and C. L. Marshall, "Review of dimethyl carbonate (DMC) manufacture and its characteristics as a fuel additive," Energy & Fuels, vol. 11, no. 1, pp. 2–29, 1997. · ·
26. F. Rivetti, U. Romano, and D. Delledone, "Dimethyl carbonate and its production technology," in Green Chemistry, P. T. Anastas and T. C. Williamson, Eds., vol. 626 of ACS Symposium Series, pp. 70–80, 1996.
27. Y. Katrib, G. Deiber, P. Mirabel et al., "Atmospheric loss processes of dimethyl and diethyl carbonate,"Journal of Atmospheric Chemistry, vol. 43, no. 3, pp. 151–174, 2002. · ·
28. Y. Yu, X. Liu, W. Zhang et al., "Electrosynthesis of dimethyl carbonate from methanol and carbon monoxide under mild conditions," Industrial and Engineering Chemistry Research, vol. 52, no. 21, pp. 6901–6907, 2013. · ·
29. M. Wang, N. Zhao, W. Wei, and Y. Sun, "Synthesis of dimethyl carbonate from urea and methanol over ZnO," Industrial and Engineering Chemistry Research, vol. 44, no. 19, pp. 7596–7599, 2005. · ·
30. W. Zhao, F. Wang, W. Peng et al., "Synthesis of dimethyl carbonate from methyl carbamate and methanol with zinc compounds as catalysts," Industrial and Engineering Chemistry Research, vol. 47, no. 16, pp. 5913–5917, 2008. · ·
31. P. Unnikrishnan and D. Srinivas, "Calcined, rare earth modified hydrotalcite as a solid, reusable catalyst for dimethyl carbonate synthesis," Industrial and Engineering Chemistry Research, vol. 51, no. 18, pp. 6356–6363, 2012. · ·

32. E. Pongrácz, E. Turpeinen, R. Raudaskoski, D. Ballivet-Tkatchenko, and R. L. Keiski, "CO_2: from waste to resource for methanol-based processes," Proceedings of Institution of Civil Engineers: Waste and Resource Management, vol. 162, no. 4, pp. 215–220, 2009.
33. F. Cavani, G. Centi, S. Perathoner, and F. Trifiro, Sustainable Industrial Process, Wiley-VCH, New York, NY, USA, 2009.
34. D. Ballivet-Tkatchenko and S. Sorokina, "Open chain organic carbonates," in Carbon Dioxide Recovery and Utilization, M. Aresta, Ed., pp. 261–277, Kluwer Academic Publishers, Dordrecht, The Netherlands, 2003.
35. S. K. Kabra, E. Turpeinen, G. D. Yadav, and R. Keiski, "Direct synthesis of dimethyl carbonate from methanol and carbon dioxide: a thermodynamic and experimental study," The Journal of Supercritical Fluids. To be submitted.
36. P. Adhuri and G. D. Yadav, Insight into catalytic green chemistry and technology of industrial relevance [M.S. thesis], University of Mumbai, 2007.
37. H. Kawanami, A. Sasaki, K. Matsui, and Y. Ikushima, "A rapid and effective synthesis of propylene carbonate using a supercritical CO_2-ionic liquid system," Chemical Communications, vol. 9, no. 7, pp. 896–897, 2003.
38. M. Nicola Di, C. Fusi, F. Rivetti, and G. Sasselli, "Patent: process for producing dimethyl carbonate," Tech. Rep. EP 0460732, EniChem Synthesis S.p.A., 1991.
39. Y. Du, F. Cai, D.-L. Kong, and L.-N. He, "Organic solvent-free process for the synthesis of propylene carbonate from supercritical carbon dioxide and propylene oxide catalyzed by insoluble ion exchange resins," Green Chemistry, vol. 7, no. 7, pp. 518–523, 2005.
40. M. Cinelli, S. R. Coles, and K. Kirwan, "Analysis of the potentials of multi criteria decision analysis methods to conduct sustainability assessment," Ecological Indicators, vol. 46, pp. 138–148, 2014.
41. V. Belton and T. J. Stewart, Multiple Criterial Decision Analysis. An Integrated Approach, Kluwer Academic, Boston, Mass, USA, 2002.

42. T. Buchholz, E. Rametsteiner, T. A. Volk, and V. A. Luzadis, "Multi Criteria Analysis for bioenergy systems assessments," Energy Policy, vol. 37, no. 2, pp. 484–495, 2009. · ·
43. L. C. Dias and A. R. Domingues, "On multi-criteria sustainability assessment: spider-gram surface and dependence biases," Applied Energy, vol. 113, pp. 159–163, 2014. · ·
44. V. Eta, P. Mäki-Arvela, A.-R. Leino et al., "Synthesis of dimethyl carbonate from methanol and carbon dioxide: circumventing thermodynamic limitations," Industrial and Engineering Chemistry Research, vol. 49, no. 20, pp. 9609–9617, 2010. · ·

Chapter 3

Improved Safety for Automotive Lithium Batteries: An Innovative Approach to include an Emergency Cooling Element

Peter Kritzer[1], Harry Döring[2], and Brita Emermacher[2]

[1]Advanced Product Engineering, Freudenberg Sealing Technologies GmbH & Co. KG, Weinheim, Germany
[2]Electrochemical Energy Technologies, Zentrum für Sonnenergie und Wasserstoff-Forschung Baden-Württemberg, Ulm, Germany

ABSTRACT

This paper describes a concept for an independent and redundant safety concept for Lithium batteries in Electric and Hybrid Electric Vehicles. This concept includes an emergency cooling system based on pressurized carbon dioxide (CO_2). Since carbon dioxide (CO_2) is a possible medium of future mobile air conditioning (MAC) systems, the MAC system can be utilized for the one-time emergency cooling described in this paper. In the first part of the paper, some major safety aspects of automotive Li batteries are highlighted. In the second section, the paper describes a technical approach, how these batteries can be made safer. Pressurized CO_2, which is a promising candidate for cooling liquids used in future mobile air conditioning (MAC) systems, is used to effectively cool down an overheating or up-heating battery in a critical state. The safety system thereby is not based on an electrical effect, but on a direct and fast-reacting thermal conduction, avoiding a thermal runaway of individual cells. The application of the proposed system is to act preventively just before the thermal runaway gets uncontrollable. In this case, the limited amount of CO_2, which is available in the MAC system, fulfils the emergency cooling requirements. The combination of standard car components for the concept leads to an only moderate increase of the total weight and the additional system costs. Therefore, the described system might be of interest for car, battery and air conditioning system producers. This paper explains that the synergetic combination of CO_2-based MAC systems and Li-based batteries is an innovative approach to improve environmental compatibility in future vehicles. The concept is proven experimentally on a lab scale with battery cells and battery packs consisting of four serially connected cells, respectively.

INTRODUCTION

The present paper highlights technical trends in two different sectors—EV/HEV batteries and mobile air conditioning (MAC) systems. It describes the synergetic combination of both trends.

The idea of this paper is based on the following facts:

- Hybrid Electric Vehicles (HEVs) will be equipped with Lithium-based batteries (Li-Ion, or Li-Polymer, resp.), which will more or less substitute the current Nickel-Metal Hydrid (NiMH) battery systems for performance reasons.
- Plug-in Hybrid Electric Vehicles (PHEVs) and Pure electric vehicles (EVs) will exclusively be equipped with Lithium batteries.
- However, safety over the complete product life cycle still is a major issue for all Li batteries.

- From 2011 on, the European legislative 2006/40/EG bans partly fluorinated saturated substances like 1,1,1,2- Tetrafluoroethane (R134a) from mobile air conditioning (MAC) systems [1] . These substances bear a huge global warming potential (GWP) of approximately 1300 (compared to CO_2), and therefore significantly contribute to the global greenhouse effect, when emitted into the environment (leakage, scrapping). Current substitutes like $CF_3CF=CH_2$ (1234yf) are currently favoured media, since they can be more or less used in current MAC systems. However, these substances show significant flammability and therefore, they may be a safety risk [2] . An alternative candidate is carbon dioxide (CO_2; R744) as a cooling agent [3] . The greenhouse potential of CO_2 is by more than three orders of magnitude lower compared to the currently used media and approximately four times lower compared that of R1234yf. In comparison to other replacement candidates, CO_2 is completely non-flammable. Furthermore, CO_2 based cooling cycles possess much higher heating efficiency at low temperatures, which directly influences the electrical driving range of EVs at low temperatures.
- The CO_2 cycle in future air conditioning systems is schematically shown in Figure 1. In the high-pressure part, the CO_2 pressure typically is rd 14 MPa (140 bar). An additional disadvantage compared to CO_2 is their higher GWP. Some characteristics of the currently used and future cooling media for MAC systems are listed in Table 1.
- CO_2 possesses a positive Joule-Thomson coefficient. This means that when CO_2 is rapidly expanded from the above-mentioned high pressure, e.g. through a suitable valve or a nozzle, it cools down drastically. Therefore, expanding CO_2 can be used as an effective cooling agent.

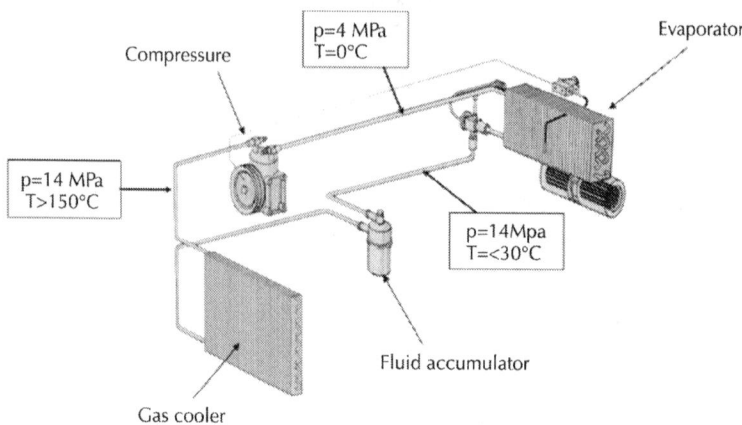

Figure 1: Schematic of a mobile air conditioning (MAC) system based on carbon dioxide (CO_2) as a cooling medium [Figure reprinted by kind permission of Freudenberg Sealing Technologies GmbH & Co. KG, Germany].

Table 1: Overview of current and future cooling media for MAC systems

	R134a	R1234yf	R1234yf + CF3I	R744
Chemical formula	CF3-CFH2	CF3CF=CH3	CF3CF=CH3/CF3I	CO2
Toxicity	no	no	yes	no
Flammability	no	yes	no	no
GWP	1300	3 - 4	3 - 4	1
Reactivity with Lithium	yes	yes	yes	no
Usable in current MAC systems	Currently standard	yes	yes	No; re-design necessary

*GWP = Global Warming Potential; normalized on CO_2.

THE SAFETY ASPECT OF LITHIUM BATTERIES

The safety aspect of Li-based batteries is well known [4] -[9] . Table 2 gives an overview for the typical requirements of Li battery systems used in consumer and automotive applications. For automotive batteries, safety requirements become much more important not only due to the higher energy amounts, but also due to the harsher environments and lifetime requirements. Thus, "safety" is the most important task for traction battery systems in automotive systems.

The present approach gives an additional, redundant tool to effectively and safely shut down a Li-ion battery in a critical, overheated state. This approach is based on an cell-external feature, so the electrochemistry itself is not influenced by this system.

The components/aggregates used for this system are already present in a future car, so the additional safety feature does not need much additional equipment. Consequently, weight increase and additional costs for the car are low to moderate

THE CONCEPT OF A REDUNDANT AND INDEPENDENT SAFETY SYSTEM

Automotive fire extinguishing equipment based on CO_2 as medium is described in patent literature [10] . It is also described to use CO_2 coming from a MAC system to extinguish a burning battery [11] . However, all these concepts are not transferable to EV applications, since the amount of CO_2 necessary to extinguish a battery-initiated fire exceeds the amount of CO_2 in an automotive MAC.

Therefore, the present concept uses the CO_2 of the MAC only to cool down the battery prior to any fire and/or explosion [12] .

Basic feature of the safety system described in this paper is a pressure tube starting in the high-pressure, CO_2 containing part of the MAC system [12] . At this point, CO_2 is in its semi-critical state (the pressure is typically 14 MPa, which is above the critical pressure of

7.3 MPa, but the temperature is just below the critical temperature of 31°C). Through this tube, CO_2 can be transported from the MAC circuit system, normally towards the HEV/EV battery. The tubing could be placed together with the electrical cables linking the battery and the electric engine. The system is schematically shown for a HEV in Figure 2.

The tube ends inside the battery housing, close to the cells. Under normal conditions, it is completely filled with semi-critical CO_2. At the outlet point, a special expansion valve is placed, which can control the rapidly expanding and cooling medium.

In case of a malfunction of the battery, this valve is opened and the highly-compressed liquid CO_2 expands through this nozzle forming gaseous or solid CO_2. Thereby, it rapidly cools down due to the positive Joule Thomson-Effect coefficient for CO_2. This cooling effect can be effectively used to cool down individual cells inside the battery. The refrigeration capacity of CO_2 amounts to 645 kJ/kg referring to a temperature increase from −79°C to 0°C.

The gas outlet system itself can be combined intelligently with the battery management system (BMS), or it can possess a redundant sensor (e.g. temperature measurement by additional thermocouples, or a pressure measurement detecting an extraordinary pressure increase in stack pressure). Ideally, it should activate, when cells reach critical values, and before a real "battery accident" occurs.

By using multiple expansion valves and tubes placed in different sections of the battery, a selective cooling of single overheating cells is possible (see Figure 3). This assembly allows the direct release of the cooling me dium precisely into these areas, where the overheating is detected, which assures an economic and effective use of the cooling medium in case of emergency.

Table 2: Typical parameters for Li batteries used in consumer and automotive

	Batteries in Consumer Applications	Batteries in Automotive Applications
Mass Battery system	<0.1 kg	>100 kg
Cell capacity	1 Ah	>20 Ah

System Energy	0.002 ... 0.02 kWh	1 kWh (HEV) > 15 kWh (EV)
System Voltage	3.6 ... 11 V	100 V
Max. Discharging Currents	<1 A	> 100 A
Max. Charging Currents	<1 A	50 A
Operating Temperature	0...40°C	−40 ... + 85°C
Further Conditions	Dust; Splash Water	Dirt, Oil, Water, Vibrations
Typical Specified Lifetime	3 years	>10 - 15 years
Typical Tolerated Scrap Rate (Cells)	0.1%	1 ppm

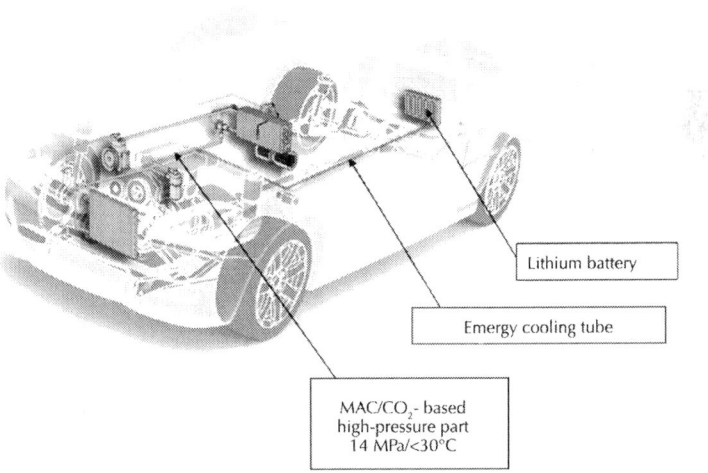

Figure 2: Concept for combining CO_2-based mobile air conditioning (MAC) systems and HEV/EV Lithium battery. In case of emergency (especially overheating of the battery), the CO_2 can be used to rapidly and effectively cool

down the battery medium [figure reprinted by kind permission of Freudenberg Sealing Technologies GmbH & Co. KG, Germany].

In order to reduce leakage through the elastomeric seals and thus to prolong service times of the MAC, a main valve may be placed in the main tube front of the branching. By using elastomers with low values CO_2 penetration [13], the long-term effectiveness of the system can be guaranteed.

A less complicated system can be generated if the pressure tubes are equipped with melting fuses at the cell end. When cell surface temperature exceeds values of approx. 80°C - 100°C, the fuses open immediately and emit the cooling media onto the surface.

EXPERIMENTAL SET UP

The concept has been tested in a lab scale at ZSW, Ulm, Germany. Battery failure was induced by a severe overcharge process. As usually, Li batteries do not have a side reaction employable for overcharge, overcharge processes are leading to a damage of the system and might be happen in a catastrophic event.

Model cell For the study, a laminated sheet cell (LSB; Kokam, KR) has been used. The nominal capacity of the cell was 4.0 Ah and the nominal voltage 3.7 V.

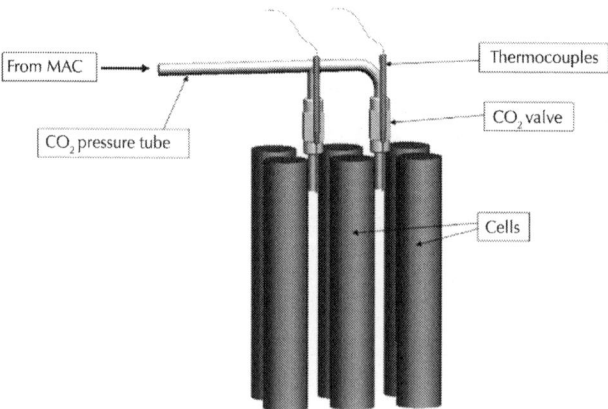

Figure 3: Schematic view of a battery system consisting of round cells equipped with thermocouples and cooling system. Each thermocouple is linked with

a CO_2 tube and a corresponding outlet valve. This arrangement allows the directed release of the cooling medium into exact these areas, where the overheating is detected. This assures the economic and effective use of the cooling medium in case of emergency [figure reprinted by kind permission of Freudenberg Sealing Technologies GmbH & Co. KG, Germany].

Tow thermocouples (type K) have been mounted to the front and the backside of the cell.

CO_2 dosage.

CO_2 came from a pressurized reservoir (p = 5.8 MPa @ 20°C) and was filled in a pressure stable tube (volume ca 170 ml). Than the main reservoir was closed, so that the total amount of the CO_2 was limited (p = 5.8 MPa, 170 ml). The outlet nozzle (diameter 2 mm) of the tube was controlled by a magnetic valve. The arrangement of the testing equipment is shown in Figure 4.

CELL OVERCHARGE (5C/8.4V/CELL) WITHOUT EMERGENCY COOLING

The cell was pre-charged to 85% SOC and was fastened to a piece of insulation material. Overcharge was at 20 a (corresponding to a charging rate of 5C). The cell inflated, opened and released electrolyte vapour. When a voltage of 8.4 V was reached (after approx. 8 min of overcharging), charging was terminated (switch off the charger).

At this state (ca. 153% SOC), temperatures at the cell surfaces were about 90°C. In the following rd. 8 min, cell surface temperature gradually increased to values of 120°C - 170°C (self-heating). Then, suddenly, the cell ruptured with flying sparks and smoke emission. The cell rupture was most probably due to a melt-down of the separator (thermal runaway occurred) and the cell burned. On the back cell surface temperatures of 485°C and on the front surface of 306°C were reached. The fire was extinguished.

The course of the experiment is shown in Figure 5 and the destroyed cell in Figure 6.

CELL OVERCHARGE (5C/8.4V/CELL) WITH EMERGENCY COOLING

The experiment was started in analogy to the previous test. After having reached the charging voltage limitation of 8.4 V, the charging was stopped at a similar point as before. The surface temperature of the cell was around 90°C. A steady surface temperature rise was observed for further 9 s after charging stopped. At that point the cell was cooled down by a 20 sec pulse of expanded CO_2 (approx. 50 g). Front Surface cell temperature immediately dropped down from 93°C to −49°C. After CO_2 pulse stop, cell surface temperature increased again to values of max. 90°C (back side) and max 78°C on the front side, but then temperature cooled down gradually. No rupture occurred.

The course of the experiment is shown in Figure 7; cell open, but not ruptured in Figure 8.

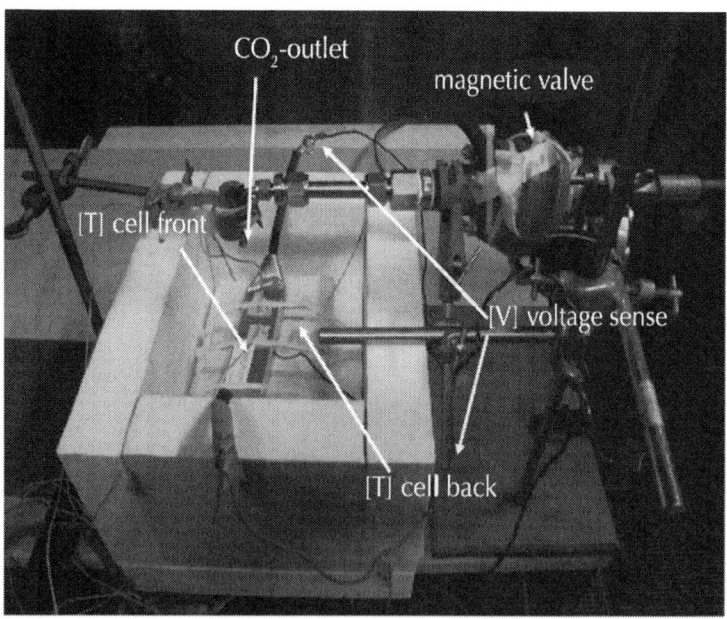

Figure 4: Test set up with LSB cell, temperature sensors and CO_2-cooling outlet.

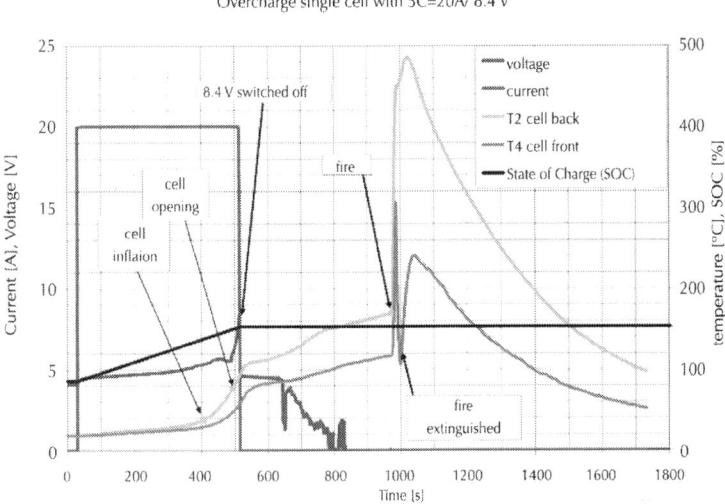

Figure 5: Course of experiment with one cell, without emergency cooling. The internal overheating of the cell finally resulted in a rupture/fire.

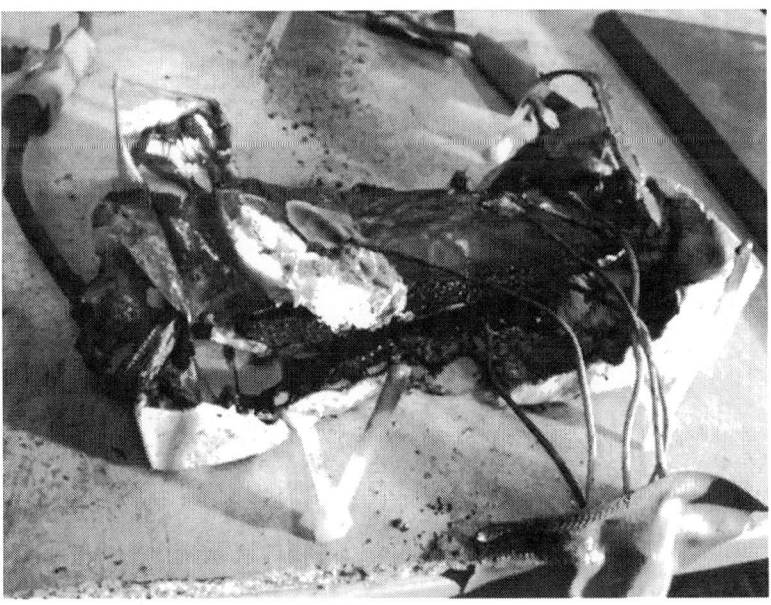

Figure 6: Destroyed overcharge cell after the test, burned and ruptured.

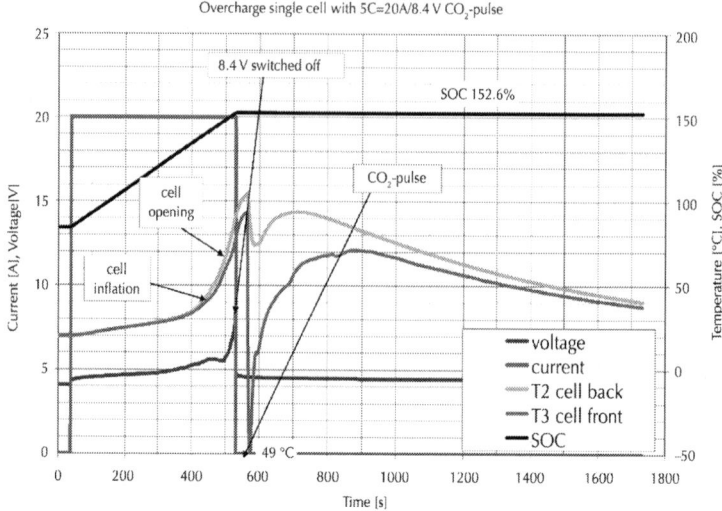

Figure 7: Course of experiment with one cell, with emergency cooling at t = 550 s. No internal overheating of the cell occurred; neither a rupture nor a fire was observed.

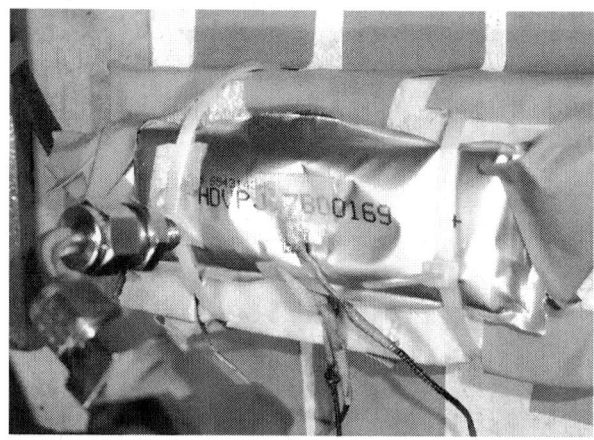

Figure 8: Cell after emergency cooling experiment. Cell inflated and the CO_2 pulse resulted in a dented cell. No cell rupture occurred.

When the cell was switched off at the same critical point, in one case, the disaster occurred. However, with a short cooling pulse through expanded CO_2, the mode could be switched over and the cell was safe.

Cell Pack Overcharge (5C) without Emergency Cooling

A battery pack of 4 cells (connected in series; side-by side-arrangement) was used for the following two experiments. The intention of the experiment is to analyse, whether a failure in a single cell is propagated to its neighbours. In this case, a failing cell would cause a catastrophic failure of the whole battery. If so, it would be important to detect the failure of this single cell as early as possible and to transfer the cell into a safe status; realising this on time, the whole battery could be protected by a minimum of action.

For this test, a particular cell (9) (Figure 9) has a significant higher SOC compared to the other 3 cells (7, 8, 10). The state of charge (SOC) of one of the inner cells was 50% while the other 3 cells had an SOC of 0%. The course of the experiment is shown in Figure 10. The pack was (over) charged with 20 A (corresponding to a charging rate of 5C). As the cell voltage of the pre-charged cell 9 could not be controlled separately, the overall voltage for the 4 series connected cells was limited to 23 V.

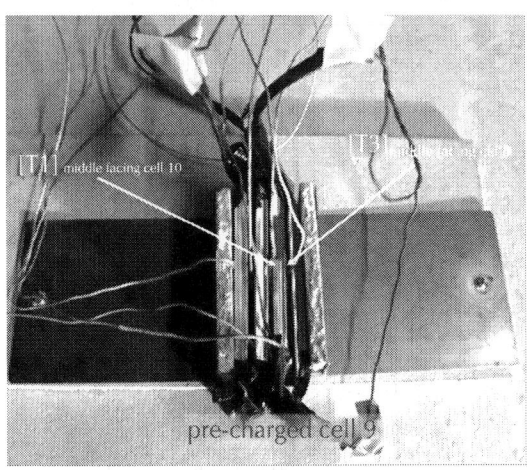

Figure 9: Test set up with pack consisting of 4 serial LSB cells, SOC of cell 9 was 50 % and for the remaining three cells 0%.

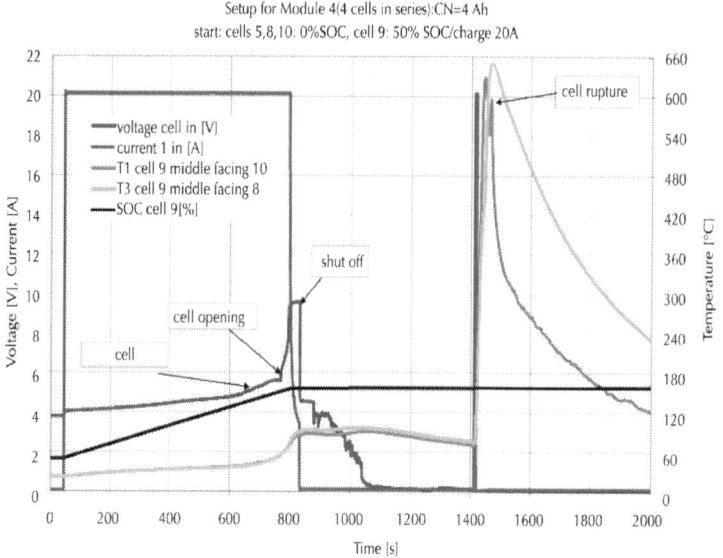

Figure 10: Course of experiment without emergency cooling. Cell failing during restart charge resulted in a rupture/fire of the whole pack.

The cell inflated, opened and released electrolyte vapour at a SOC between 140 to 150%. When the overall voltage limit of 23 V was reached (after 830 seconds), charging was terminated. Cell voltage dropped to 4.5 V and decreased in the following. Further increase in temperature was observed, however, to a lower extent compared to the single cell behaviour. This might be related to the additional heat transfer to the cell neighbours, which are not overcharged yet. This allows the overcharged cell to steady, and the cell is cooling down slowly, without the expected cell rupture and fire.

However, when switching on the charger again, the cell caught fire immediately and ruptured. The fire was transferred to the neighbouring cells and the whole module was damaged as shown inFigure 11. As anticipated, not only the overcharged cell was damaged, the failure spread to the other cells as well.

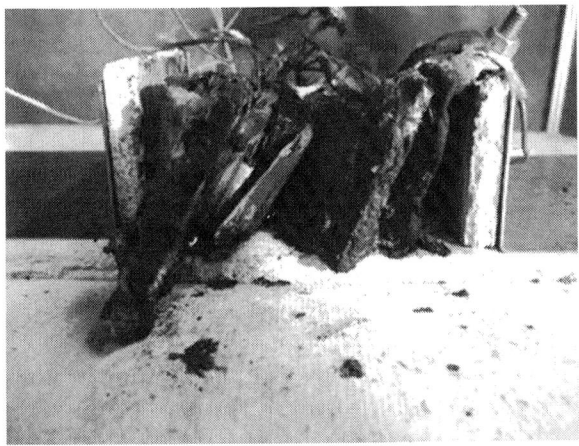

Figure 11: Module of 4 cells after overcharge of a single cell.

Cell Pack Overcharge (5C) with Emergency Cooling

The setup corresponds to experiment C. The prepared cell 9 was pre-charged to 90% SOC, while the other cells were charged to 50% SOC. The course of the experiment is shown in Figure 12. The pack was (over)charged with 20 A (corresponding to a charging rate of 5C).

Cell 9 inflated, opened and released electrolyte. When module voltage reached 23V the charging current was stopped. The CO_2-pulse was immediately released after the module charging had been stopped. The temperature dropped down by the CO_2 pulse and recovered to a level about 20°K below the previous temperature. Afterwards, a slow continuous decrease in temperature was observed and the cell voltage kept at around 4.5 V. Neither the cell nor the module was ignited and no cell rupture was observed.

DISCUSSION OF THE RESULTS

The above experiments show, that a high power Lithium battery in a critical, overcharged state can effectively be switched down to a safe level by the described emergency cooling concept.

Parallel experiments without the emergency cooling resulted in rupture and fire of the cells.

A "scale-up" of the concept to a small pack consisting of four serial cells showed similar results.

In both cases, the amount of the emergency cooling medium CO_2 can be adjusted to a low level, so the amount of cooling medium coming from a MAC system (rd. 500 g) is sufficient to economically cool down cells in an automotive Li battery from a critical to a safe. This amount might not be high enough to extinguish the battery once it is burning.

For the tests, a very high voltage limit was used for triggering the emergency cooling. However, the trigger for initiating the emergency cooling might be as well the temperature measured with a thermocouple at the cell surface or any other signals (e.g. stack pressure). In later automotive applications, the sensor signal might be connected to the battery management system, and the occurrence of different signals and their combination might be used as a trigger signal, avoiding faulty activation. Completely stand-alone arrangements might be possible as well. Such can be fusible cut-outs at the ends of the tube outlets, melting and opening when temperature exceeds some 90°C - 100°C and thus release the pressurized CO_2. (Figure 13)

CONCLUSIONS

This paper describes a concept increasing safety of Hybrid Electric Vehicles (HEVs) and Electric Vehicles (EVs) equipped with Lithium-based batteries. Further battery sealing-relevant products are described elsewhere [14].

Currently, automotive Li batteries are equipped with a State of Charge (SoC) sensor on cell basis. Thus, an automotive battery system requires some 100 - 200 individual wires, which are contacted to the power electronic system.

This raises two questions in this content:

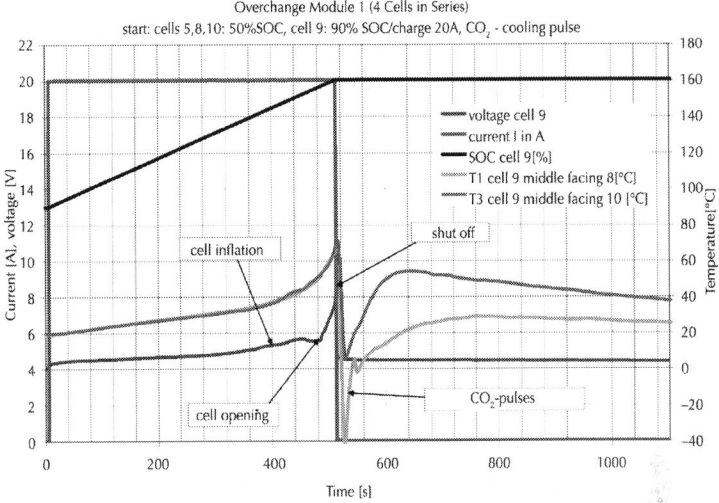

Figure 12: Course of experiment with a battery pack, with emergency cooling no internal overheating of the pack occurred; neither a rupture nor a fire was observed.

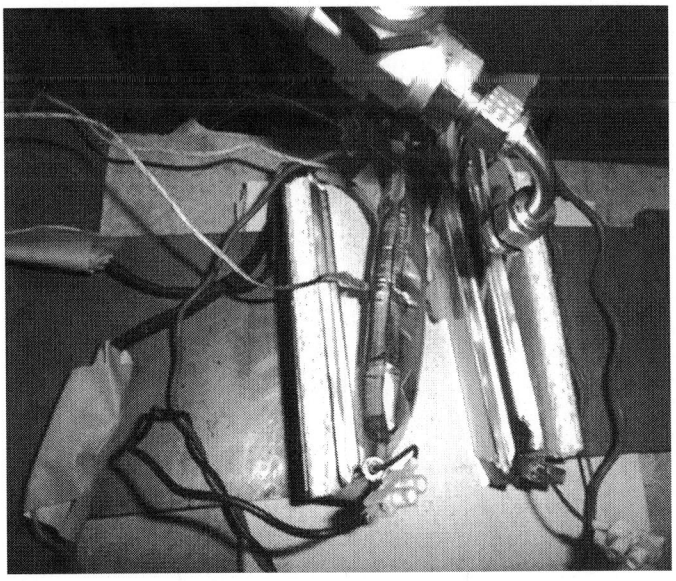

Figure 13: Cell 9 inflated / module with CO_2-cooling.

- What happens, if the electrical contact of one of the tiny sensor wires is opened/destroyed? If so, will the driver of an EV immediately be informed in this case?
- Will future battery systems be equipped with a sensor system on cell basis?

If one of the questions is answered with "no", a single cell overcharging with high charging currents cannot be excluded. In such a case, a critical, overheating state as it is simulated in the present article cannot be excluded.

The additional safety described here can help to avoid the final consequences of battery fire/explosion. The safety system consists of a synergetic combination of a CO_2-containing mobile air conditioning (MAC) system and the Li-based battery system. In case of cell overheating, the compressed CO_2 inside the MAC system can be expanded into the battery leading to a rapid and effective cool-down due to its positive Joule-Thomson coefficient. It must be clearly pointed out that the focus of the concept is to cool down parts of the battery selectively in order to prevent an ignition. The amount of CO_2 in a MAC system is by far not sufficient to extinguish an already existing fire.

The concept has been proven in lab-scale experiments in this paper. It has been shown that a high power Lithium battery in a critical, overcharged state can effectively be switched down to a safe level by the described emergency cooling concept. Parallel experiments without the emergency cooling result in rupture and fire of the cells. A "scale-up" of the concept to a small pack consisting of four serial cells shows similar results.

From the point of view of a car producer, the concept should be attractive, since additional costs and weight of this redundant safety system are moderate. On the other hand, this additional safety effect helps to support the implementation of the energy-efficient CO_2-containing MAC systems for automotive applications.

Therefore, the synergetic combination of CO_2-based MAC systems and Li-based automotive batteries can be seen as an innovative approach to improve environmental compatibility of future vehicles.

REFERENCES

1. European Parlament Regulation: RICHTLINIE 2006/40/EG des Europäischen Parlaments und des Rates vom 17. Mai 2006 üBer Emissionen aus Klimaanlagen in Kraftfahrzeugen und zur Änderung der Richtlinie 70/156/EWG des Rates; Amtsblatt der Europäischen Union 14.06.2006, L 161/12-L 161/18.
2. Graz, M. and Wuitz, U. (2008) Flammability Investigation of Different Refrigerants Using an Operating MAC System in a Simulated Front End Collision Situation. www.r744.com
3. Morgenstern, C. (2008) R744 MAC Status and System Standardisation. VDA Winter Meeting, Saalfelden.
4. Roth, E.P., Doughty, D.H. and Pile, D.L. (2007) Effects of Separator Breakdown on Abuse Response of 18650 Li-Ion Cells. Journal of Power Sources, 174, 579-583.http://dx.doi.org/10.1016/j.jpowsour.2007.06.163
5. Doh, C.H., Kim, D.H., Kim, H.S., Shin, H.M., Jeong, Y.D., Moon, S.I., Jin, B.S., Eom, S.W., Kim, H.S., Kim, K.W., Oh, D.H. and Veluchamy, A. (2008) Veluchamy: Thermal and Electrochemical Behaviour of C/Li_xCoO_2 Cell During Safety Test. Journal of Power Sources, 175, 881-885. http://dx.doi.org/10.1016/j.jpowsour.2007.09.102
6. Balakrishnan, P.G., Ramesh, R. and Kumar, T.P. (2006) Safety Mechanisms in Lithium-Ion Batteries. Journal of Power Sources, 155, 401-414.http://dx.doi.org/10.1016/j.jpowsour.2005.12.002
7. Arora, P. and Zhang, Z. (2004) Battery Separators. Chemical Reviews, 104, 4419-4462.http://dx.doi.org/10.1021/cr020738u
8. Zhang, S.S. (2007) A Review on the Separators of Liquid Electrolyte Li-Ion Batteries. Journal of Power Sources, 164, 351-364.http://dx.doi.org/10.1016/j.jpowsour.2006.10.065
9. Zhang, S.S. (2006) A Review on Electrolyte Additives for Lithium-Ion Batteries. Journal of Power Sources, 162, 1379-1394. http://dx.doi.org/10.1016/j.jpowsour.2006.07.074
10. Ostlyngen, T.W. and Thrones, B. (1996) Method and Apparatus for Detection and Prevention of Fire Hazard. International Patent WO 96/16699.

11. Wertenbach, J. and Albersfelder, G. (1995) Kraftfahrzeug Mit Einer Klimaanlage, European Patent Application EP 0 675 013 A1.
12. Kritzer, P. and Raida, H.J. (2011) Method for Cooling of an Energy Storage Device. European Patent EP 2 045 852 B1.
13. Frenzel, U., Weiss, R. and Peterseim, V. (2008) Für saubere Luft- Dichtungsloesungen Mit Geringer CO_2-Permeation Fuer R744-Klimaanlagen. Automobilkonstruktion, 1, 42-43.

Chapter 4

Process for the Obtention of Coumaric Acid from Coumarin: Analysis of the Reaction Conditions

Néstor N. López-Castillo[1], Alma D. Rojas-Rodríguez[2], Brenda M. Porta[1], and M. Javier Cruz-Gómez[1]

[1]Chemical Engineering Department of the Faculty of Chemistry, Universidad Nacional Autónoma de México (UNAM) Coyoacan, D. F. México, México

[2]Faculty of Engineering, Universidad Anáhuac México Norte, D. F. México, México

ABSTRACT

Coumaric acid can be obtained from basic hydrolysis of coumarin, through a reaction process consisting on opening the lactone ring and cis-trans isomerization. Parameters such as reaction time, temperature, NaOH concentration, solvent and reaction atmosphere, have been thoroughly studied and analyzed, in order to determine the appropriate conditions for the maximum conversion efficiency of coumarin into coumaric acid. Experimental results show that the best conditions are a 1 hour reaction time, at 160°C, with a 20% sodium hydroxide aqueous solution, and in an inert reaction atmosphere.

INTRODUCTION

Coumarin, a source for coumaric acid, was identified in 1820 and 1868, was synthesized in a laboratory for the first [1]. It is a pleasant smelling compound which gives a characteristic odor to hay. Other simple coumarins also possess characteristic smells sometimes exploited in perfumery [2].

Coumarin derivatives also have diverse biological properties, such as enzyme inhibition, hypotoxicity, as well as, carcinogenic, anticoagulant or antibiotic action [3]. Also, some are used as dyes given their efficient light emission properties, high stability, and ease of synthesis [4].

The bicyclic ring system of chromenes, like coumarin, has inspired a number of different synthetic approaches [5, 6].

The purpose of this research was to determine the best process conditions (time, temperature, NaOH concentration and solvent, and reaction atmosphere) to obtain coumaric acid from basic hydrolysis of coumarin.

Several processes have already been developed for the conversion of coumarin into coumaric acid. Scheme 1 shows one of these methods; where coumarin (1) converts into o-coumaric acid (2) via reflux with sodium ethoxide during four hours; then, the mixture is diluted with water, removing most of the solvent with vacuum, and neutralizing the residue with concentrated hydrochloric acid [7, 8].

Other method (Scheme 2) establishes that strong nucleophiles, such as hydroxide ions, open the coumarin ring as a result of an attack at C-2 [1, 9]. Initially, the double bond is cis, this product is known as coumarinic acid (3). Acidification of the solution at this stage is followed by rapid cyclization to reform coumarin (even under slightly basic conditions). However, on prolonged contact with the base, the cis-acid slowly isomerizes into trans-isomer (4) (known as coumaric acid). Acidification at this step allows isolating free hydroxy-acid.

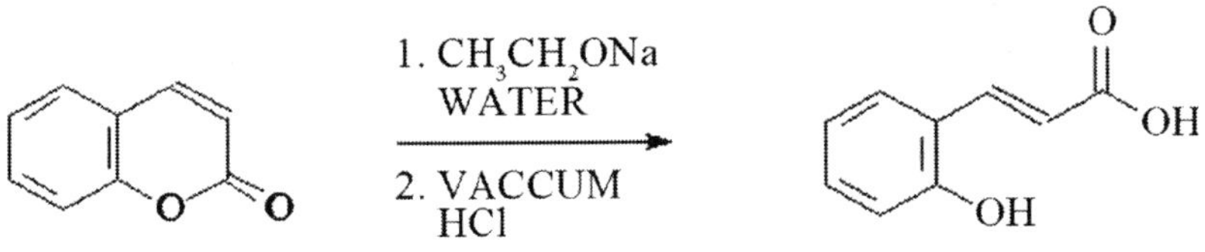

Scheme 1: Reaction to obtain coumaric acid from coumarin.

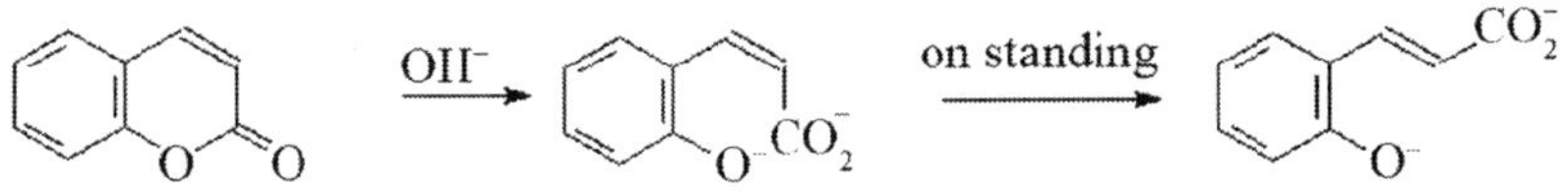

Scheme 2: Reaction to obtain coumaric acid from coumarin with strong nucleophiles.

Other authors point out that when coumarin is treated with ethanolic sodium ethoxide at reflux, the lactone ring is opened (Scheme 3) to give 2-(ethylpropenoate)-phenoxide (5). Subsequent treatment of (5) with water and evaporation of the solvent brought about saponification of the ester and, with careful acidification, produces ocoumaric acid [10-12].

Other technique [13-15] establishes that when coumarin is heated with a sodium bisulfite solution, it dissolves completely and a sulfonate compound (6) is formed (Scheme 4), which contains the lactone ring intact. Alkali splits off the sulfo group, liberating coumarin, which recombines with neutral sulfite and forms a hydrocoumaric acid derivative. The hydrocoumaric sodium sulfonate (7) is treated with 50% potassium hydroxide solution and the mixture is evaporated to dryness in a water bath. The residue is then dissolved in water, and the cooled solution is acidified with hydrochloric acid. The precipitate obtained is coumaric acid.

MATERIALS AND METHODS

Reaction

Coumarin (0.137 mol) was dissolved in a sodium hydroxide solution on a PARR reactor. The reactor was purged for about 15 to 20 minutes, to maintain the required reaction atmosphere.

When the reaction was finished, the reaction mixture was cooled into a flask, which was set into an ice bath to reach 4°C. Afterwards, concentrated hydrochloric acid was slowly added, until a white precipitate appeared at pH 5.

The precipitate, mostly coumaric acid, was filtrated and washed with cold water. Then, more hydrochloric acid was added to the liquid phase until a second precipitate appeared. Used reagents like H_2O, HCl, and NaOH were of AR grade.

Purification

The precipitate, containing coumaric acid and coumarin, was placed in a solid-liquid extraction system (Soxhlet) with chloroform. Coumarin was extracted and coumaric acid remained in the extraction thimble.

Analysis Methods

The chemical composition of the products was determined by Gas chromatography. The sample was heated from 50°C up to 100°C at 20°C/min, then kept at 100°C for 3 minutes and heated again to 300°C at the same heating rate. Afterwards, the sample was kept at 300°C, for 3.5 minutes. The equipment was flushed with helium at 2.0 mL/min and at pressure of 15.5 psi.

RESULTS AND DISCUSSION

Effect of Reaction Atmosphere

The effect of reaction atmosphere was studied for coumarin reaction with 20% of NaOH on water at constant temperature of 160°C and 1 h time period.

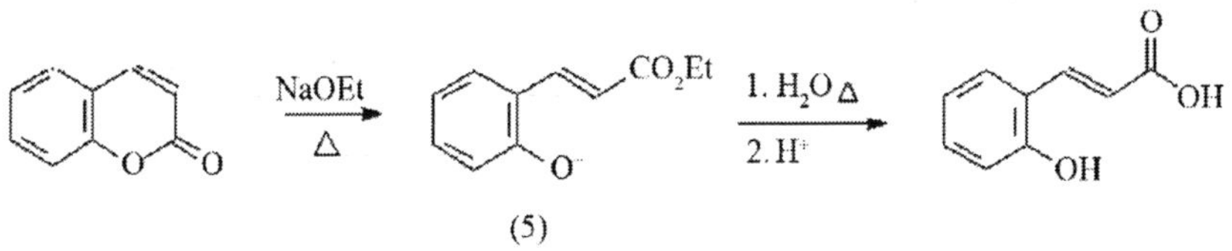

Scheme 3: Reaction to obtain coumaric acid from coumarin using ethanolic sodium ethoxide.

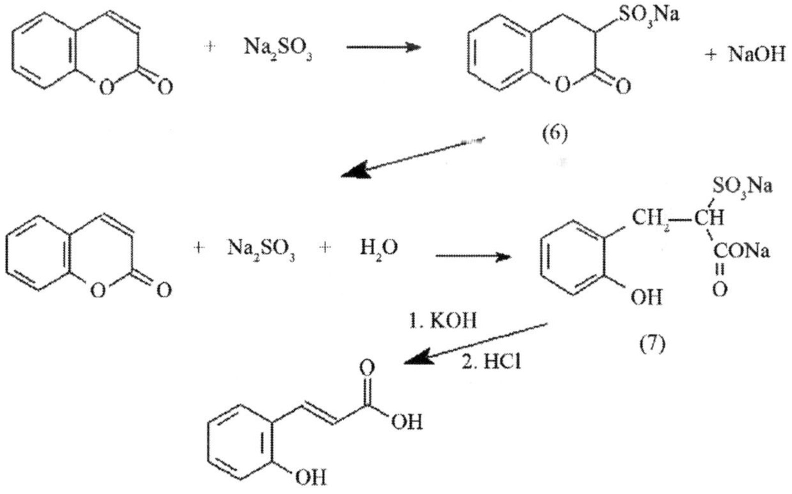

Scheme 4: Reaction to obtain coumaric acid from coumarin using sodium bisulfite.

Figure 1 shows the necessity of using an inert atmosphere. When the reaction is carried out without eliminating air from the medium, the mixture favors oxidation reactions to give by-products, which complicate the handling of material and the yield decreases (Table 1). Therefore, it was decided to pass an inert gas for 20 minutes to eliminate the air from the medium.

Effect of the Reaction Temperature

The effect of the reaction temperature on the conversion of coumarin into coumaric acid is displayed in Figure 2, for 1 h time period, 20% NaOH aqueous solution, and in a helium atmosphere. The amount of reaction yield is shown in Table 2.

Table 1: Effect of the reaction atmosphere

Reaction atmosphere	Helium	Nitrogen	Air
Yield [%]	71.7	59.0	22.8

Table 2: Effect of the reaction temperature

Reaction temperature [°C]	140	160	180	200
Yield [%]	21.3	71.7	53.1	55.0

Below 120°C, there is not conversion of coumarin into coumaric acid, and above 180°C by-products are obtained (Figure 3). The peaks found at 5.77 min and 6.21 min correspond to phenol and 2-hydroxybenzaldehyde, respectively. These two compounds are due to the oxidation of the product, and the presence of them can be observed when the final reaction mixture has an orange color instead of a yellow one. When they are contained in the final reaction mixture, acidification gives a new tacky material, which complicates the handling of the product, as well as the purification process. The temperature at which the best yield is obtained is 160°C.

Effect of the Sodium Hydroxide Concentration

The effect of NaOH concentration on coumaric acid production was studied using different concentrations within the range of 10% - 25%, at a constant temperature of 160°C, and in a helium atmosphere. Results, shown in Figure 4 and Table 3, reveal that the best yield of coumaric acid was obtained at 20 wt. % of NaOH in the reaction mixture. An increase in the caustic soda content produces a variation in the yield, since by-products such as phenol may appear and complicate the purification process. Thus, 20 wt. % sodium hydroxide solution additions are considered to be the optimum concentration.

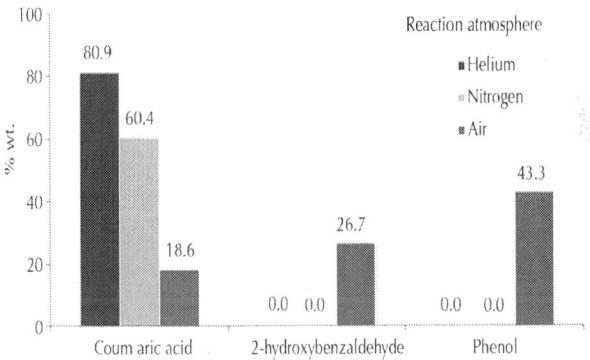

Figure 1: Effect of the reaction atmosphere.

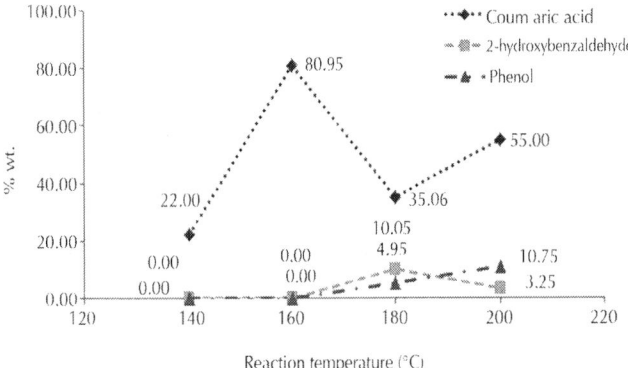

Figure 2: Effect of the reaction temperature.

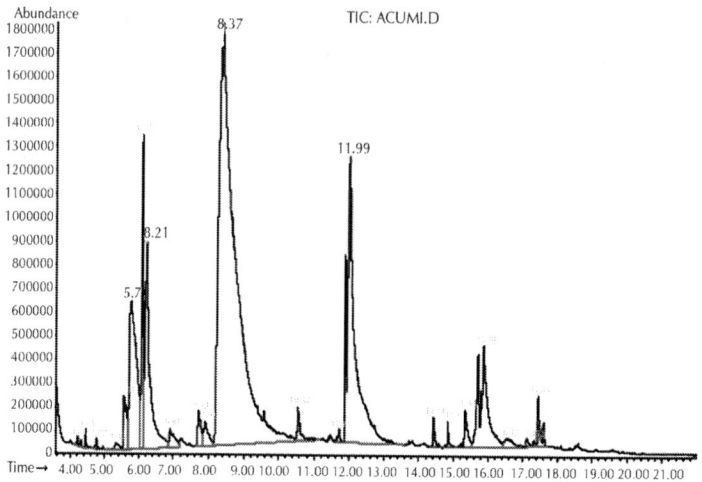

Figure 3: Chromatogram of by-products obtained when the reaction was carried out at temperatures above 180°C.

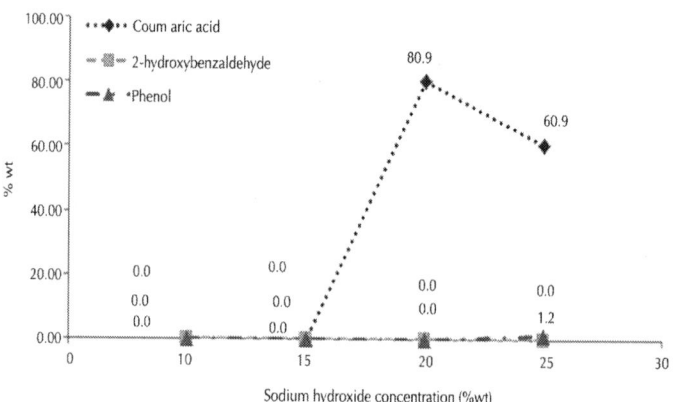

Figure 4: Effect of the sodium hydroxide concentration.

Table 3: Effect of the sodium hydroxide concentration

Sodium hydroxide concentration [%wt.]	10	15	20	25
Yield [%]	0.0	0.0	71.7	55.0

Effect of the Reaction Time

The reaction of coumarin was carried out with 20% of NaOH aqueous solution, at a constant temperature of 160°C, in a helium atmosphere, and during a period of time from 0.5 to 1.5 h.

Table 4 shows that appropriate time to obtain the best yield is 1 hour. Below one hour the yield decreases and above it, oxidation and esterification reactions are favored, thus, by-products such as benzylbenzoate and benzylcinnamate may appear (Figure 5). This can be observed in Figure 6, at the 13.3 min and 14.94 min peaks, respectively.

Table 4: Effect of the reaction time

Reaction time [h]	0.5	1.0	1.5
Yield [%]	61.2	71.7	42.9

Effect of the Solvent

The different solvents used as the reaction media are shown in Table 5. When alcohols are used as solvent, the production of coumaric acid is very low, because the group -OH may cause interference with the reaction (Figure 7). Hence, the best solvent fot this reaction is water.

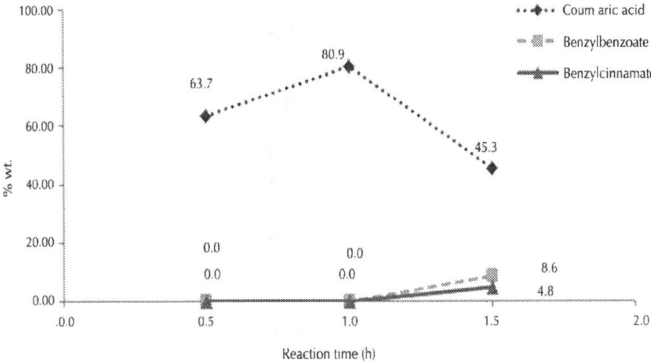

Figure 5: Effect of the reaction time.

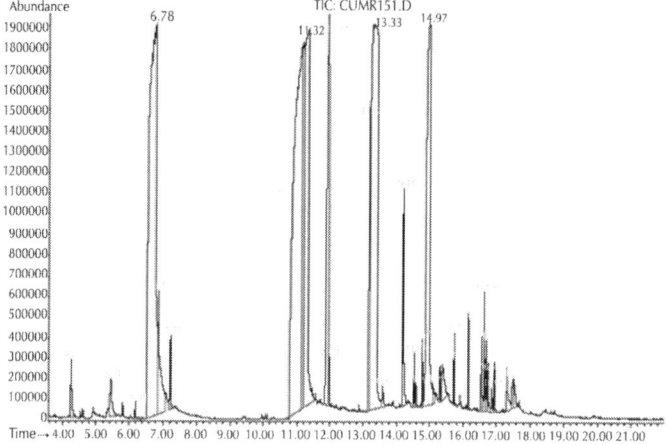

Figure 6: Chromatogram of by-products obtained when the reaction was carried out at a reaction time greater than 1 hour.

Table 5: Effect of the solvents

Solvent	Water	Methanol	Ethanol
Yield [%]	71.7	14.2	0.1

Characterization and Identification

At the optimal reactions conditions, coumaric acid can be obtained by opening the lactone ring and cis-trans isomerization. The resulting chromatograms are shown in Figure 8, where the presence of coumaric acid (8.42 min) and coumarin (12.11 min) can be observed.

Once the extraction is finished, the coumaric acid obtained had 97% purity. Figure 9 also shows the coumaric acid standard from Aldrich (b). The melting point of coumaric acid is 208°C with decomposition.

CONCLUSIONS

Methods found in literature for the production of coumaric acid may lead to the formation of raw materials or undesired products; however, it has been proven that coumaric acid can be obtained from coumarin with the conditions established in this paper.

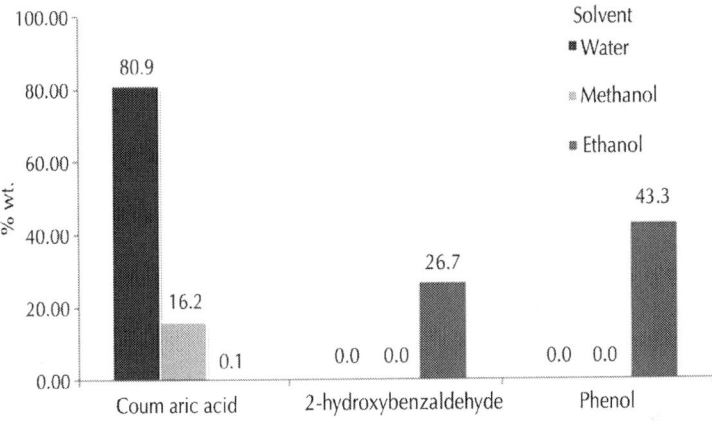

Figure 7: Effect of the solvents.

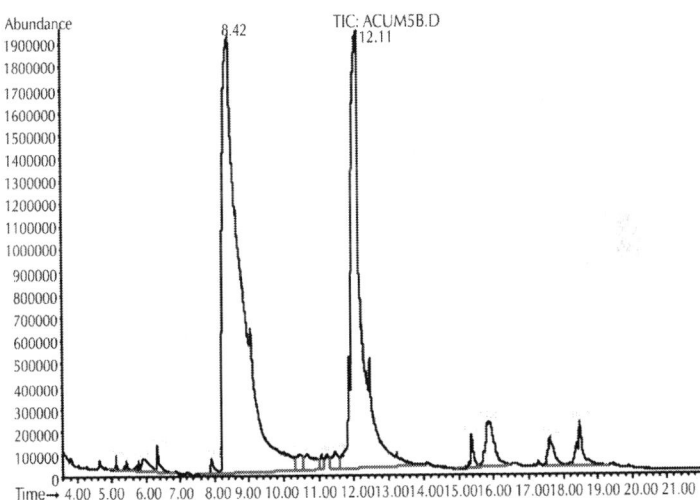

Figure 8: Chromatogram of reaction products.

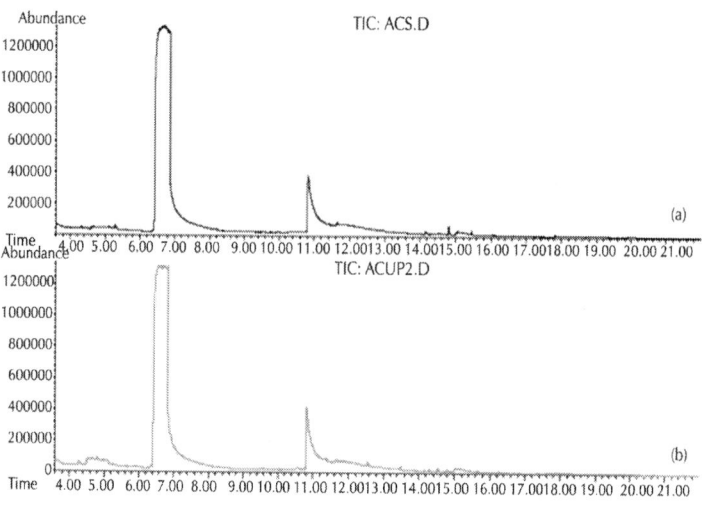

Figure 9: Chromatogram of coumaric acid obtained with the optimal reaction conditions (a) compared to the coumaric acid standard from Aldrich (b).

Under the selected experimental conditions, i.e., temperature of 160°C, reaction time of 1 h, with 20% sodium hydroxide aqueous solution, and carrying out the reaction in an inert atmosphere; coumaric acid is produced from coumarin when the lactone ring is opened through basic hydrolysis, as long as the double bond changes from cis to trans.

By applying the above mentioned parameters, the reaction yield is 70%, which represents an improvement over other methods proposed in literature. Furthermore, these conditions allow the ring opening and prevent the formation of by-products such as phenol, 2-hydroxybenzaldehyde, benzylbenzoate and benzylcinnamate.

ACKNOWLEDGEMENTS

The authors wish to acknowledge the help of Unidad de Servicio de Apoyo a la Investigación (USAI), Facultad de Química, UNAM, México, for their technical support.

REFERENCES

1. R. E. Kirk and D. F. Othmer, "Encyclopedia of Chemical Technology," 5th Edition; John Wiley & Sons Ltd., Chichester, 2007.
2. J. R. S. Hoult and M. Paydt, "Pharmacological and Biochemical Actions of Simple Coumarins: Natural Products with Therapeutic Potential," General Pharmacology, Vol. 27, No. 4, 1996, pp. 713-722. doi:10.1016/0306-3623(95)02112-4
3. E. M. Agostinha, R. Matos, C. C. S. Sousa, et al., "Energetics of Coumarin and Chromone" The Journal of Physical Chemistry B, Vol. 113, No. 32, 2009, pp. 11216- 11221.doi:10.1021/jp9026942
4. J. Preat, D. Jacquemin, et al., "Theoretical Investigation of Substituted Anthraquinone Dyes," Journal of Chemical Physics, Vol. 415, No. 20, 2005, pp. 1736-1743.doi:10.1063/1.1764497
5. B. A. Chauder, C. C. Lopes, R. S. C. Lopes, A. J. M. da Silva and V. Snieckus, "Phenylboronic Acid-Mediated Synthesis of 2H-Chromenes," Synthesis, Vol. 1998, No. 3, 1998, pp. 279-282. doi:10.1055/s-1998-2042
6. K. A. Parker and T. L. Mindt, "Electrocyclic Ring Closure of the Enols of Vinyl Quinones. A 2H-Chromene Synthesis," Organic Letters, Vol. 3 No. 24, 2001, pp. 3875-3878.doi:10.1021/ol0167199
7. I. H. Updergraff and H. G. Cassidy, "Electron Exchange—Polymer II. Vinylhydroquinone Monomer and Polymer," Journal of the American Chemical Society, Vol. 71, No. 407, 1949, pp. 407-410. doi:10.1021/ja01170a010
8. S. Patai, "The Chemistry of Functional Groups. The Chemistry of Quinoid Compounds. Part 2," 1st Edition, John Wiley & Sons Ltd., Chichester, 1974.
9. D. Burton and D. Ollis "Comprehensive Organic Chemistry. (The Synthesis and Reactions of Organic Compounds)," 1st Edition, Pergamon Press, Oxford, 1979.
10. R Murray, J Mendez and S. Brown, "The Natural Coumarins," John Wiley & Sons, Chichester, 1997.

11. E. Cingolani, "Idrolisi del Ciclo Cumarinico e Transformazione cis—Trans Degli Acidi O-Ossicinnamici," Gazzeta, Vol. 89, No. 5, 1959, pp. 195-1998.
12. E. Cingolani, "Trasformazione Degli Acidi Cumárico (Trans O-Ossicinnamici) Nelle Cumarina Corrispondenti," Gazzetta, Vol. 46, No. 9, 1959, pp. 999-1008.
13. D. Bihari and K. Krishna, "Action of Sodium Sulphite on Coumarins," Journal of the American Chemical Society, Vol. 46, No. 3, 1924, pp. 554-564.
14. R. Adams and T. E. Bockstahler, "Preparation and Reactions of O-Hydroxycinnamic Acids and Esters," Journal of the American Chemical Society, Vol. 74, No. 21, 1952, pp. 5346-5348. doi:10.1021/ja01141a038
15. F. D. Dodge, "Some Derivatives of Coumarin," Journal of the American Chemical Society, Vol. 38, No. 2, 1916, pp. 446-457. doi:10.1021/ja02259a030

Chapter 5

CO_2 Utilization: A Process Systems Engineering Vision

Ofélia de Queiroz F. Araújo[1], José Luiz de Medeiros[1] and Rita Maria B. Alves[2,3]

[1]Federal University of Rio de Janeiro, Brasil
[2]BRASKEM S.A., Brasil
[3]University of São Paulo, Brasil

INTRODUCTION

The development of economies results in increased energy consumptions, as observed in Figure 1. In the coming decades, the supply of such expanding demand will remain based on fossil fuels technologies. Expanding the share of renewable energy (e.g., biofuels in the case of transportation fuels) would require massive investments in creating a new infrastructure, which would eventually raise the standards to a new economic order entirely based on renewables in the near future. On the other hand, the current scenario involves the announcement of large proven reserves of non-conventional gas and oil and expansion of installed infrastructure of production and refining.

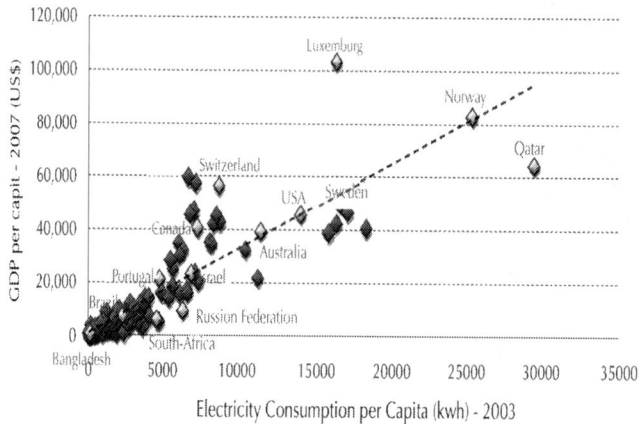

Figure 1: Economic development and electricity consumption. [graph constructed with data available at hdr.undp.org].

Fossil fuel based energy is recognized as carbon emitter. The challenge of the century is thus to expand energy supply in a carbon-constrained economy. According to the World Economic Forum (King, 2010), no truly low-carbon technology will be able to penetrate the mass market in the short term. The use of installed fossil processing infrastructure, with co-processing of biomass and fossil feedstock and capturing and utilization of emitted CO_2 is the "escape route" for a moderate transition from the present to a long-term sustainable future. In this context, putting a price on carbon will gradually build the road to a greener tomorrow. Meanwhile, bio-based products are a realistic supplement to fossil-based products, but they cannot mitigate the rising demand for fossil fuels.

According to an IEA Technology Roadmap, the manufacture of only 18 chemicals account for 80% of the energy demand in the chemical industry and 75% of its greenhouse gas (GHG) emissions. The study concludes that, "in the short to medium term (to 2025), steady progress in implementing incremental improvements and deploying best practice technologies (BPT) could provide substantial energy savings and emissions reductions compared to business as usual". "A step change in the sector's energy consumption and GHG emissions would require the development of 'game changer' technologies, such as sustainable biomass feedstocks and hydrogen from renewable energy sources which have not yet reached commercial maturity." (IEA, 2013)

With this prospect, CO_2 utilization in the short term should allow parallel production routes based on BPT, driven by emission-capture-utilization synergies. In this sense, production and conversion of synthesis gas (SYNGAS) exhibits the highest potential for medium term of commercial success. Nevertheless, it is worth noting that, while the utilization of CO_2 has potential to reduce greenhouse gas emissions into the atmosphere, CO_2 has disadvantages as a chemical reactant due to its relative significant chemical inertness. This inertness is the underlying reason why CO_2 has broad industrial application as solvent (supercritical CO_2), as fire and pest extinguisher, and as a non-toxic amusement additive in the food industry. From the standpoint of building a low-carbon economy, each potential use of CO_2 as reactant has a customarily high energy requirement, entailing associated energy-related GHG emissions. Obviously, such GHG emissions should not exceed the yield of chemical conversion of CO_2. Reverse water gas shift (RWGS) and dry reforming to yield SYNGAS and CO_2 hydrogenation to methanol are the most prominent CO_2 conversion alternatives to high volume chemical commodities. On the other hand, it is reasonable to expect that CO_2 conversion and biomass-based processing alternatives will coexist for a while with fossil-based technologies. Figure 2 illustrates the concept of co-existing fossil and biomass feedstock, having SYNGAS generation as the integration phase. In this scenario, SYNGAS can be obtained from coexisting fossil and green feedstock via gasification (of biomass, coal and heavy residues), steam reform and dry (CO_2) reform of natural gas, which, in turn, will also coexist with downstream nonconventional conversion routes – e.g., Fischer-Tropsch (FT) and methanol to olefins (MTO) – to fuels and chemical intermediates to supply the installed petrochemical industries.

In countries with large bioresources as well as significant oil and natural gas production infrastructure (e.g., USA, Canada, Mexico, China, Russia, Australia, Argentina and Brazil), fossil feedstock (mainly nonconventional oil and gas) will present the greatest challenge to biomass. In Brazil, for instance, gas supply capacity in Santos Basin increased from 600 thousand m³/d, in 2009, to 22.2 million m³/d in 2013 (PETROBRAS, 2013). The identified risks associated to such expansion are, beside depletion of natural resources, increased CO_2 emissions.

Therefore, CO_2 capture, transportation and utilization must be included in the scene since CO_2 stands conceptually as a renewable feedstock. In this sense, Figure 3 illustrates CO_2 emissions capture and utilization associated to the production and refining of fossil fuel. It is worth noting that, in addition to the variety of alternative technologies, other factors influence the conception of capture and utilization of CO_2 for the production of chemicals.

Rostrup-Nielsen and Christiansen (2011) present four recognizable trends in next generation successful plants for producing chemical commodities: (a) location of cheap raw materials; (b) economy of scale; (c) highly integrated process plants; and (d) low CO_2 footprint (t CO_2/t product) Consequently, the production chain and the associated GHG emissions are complex and require a system view for optimal decision making on technologies for production of energy and chemicals taking into account the CO_2 perspective.

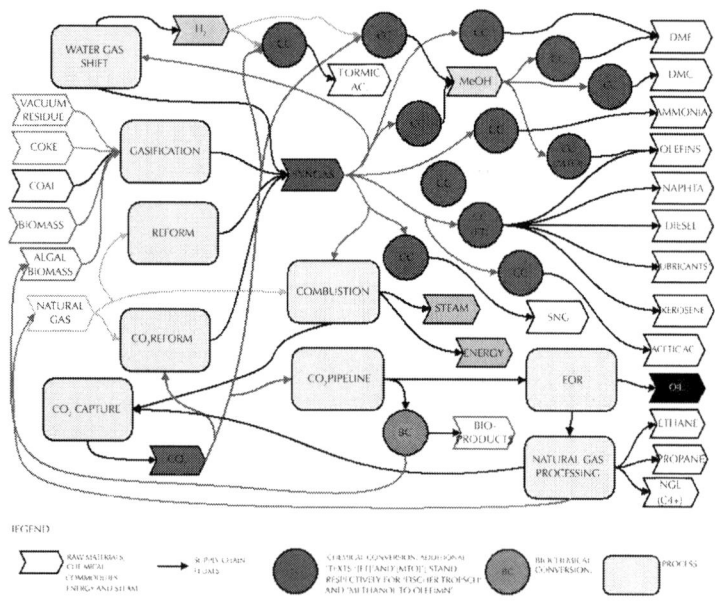

Figure 2: Coexistence of fossil and biomass feedstock uses: CO_2 capture and utilization dropped into existing production infrastructure.

Chemical Process Systems Engineering (cPSE) deals with the set of basic unit operations involved in turning raw materials into products via chemical and bio-chemical processes. In a wider definition, PSE is concerned with the optimization of decision-making process for creation and operation of chemical supply chains. Such integrated framework embraces product and process design, manufacture and distribution of chemical products with multiple and conflicting technical, economic, energetic, environmental and social objectives.

With this vision, c-PSE of industrial use and reuse of CO_2 supply chain is assessed in a life cycle approach of technological alternatives involving integration of CO_2 capture (separation), CO_2 transportation and CO_2 chemical/biochemical utilization. Apart the conventional utilizations of CO_2 listed above, there are three new potential chains of CO_2 utilizations, namely: (i) production of chemicals via chemical conversion (CC); (ii) enhanced oil recovery (EOR) and carbon geological storage (CGS); and (iii) conversion via algae and microalgae to biomass or biochemical conversion (BC).

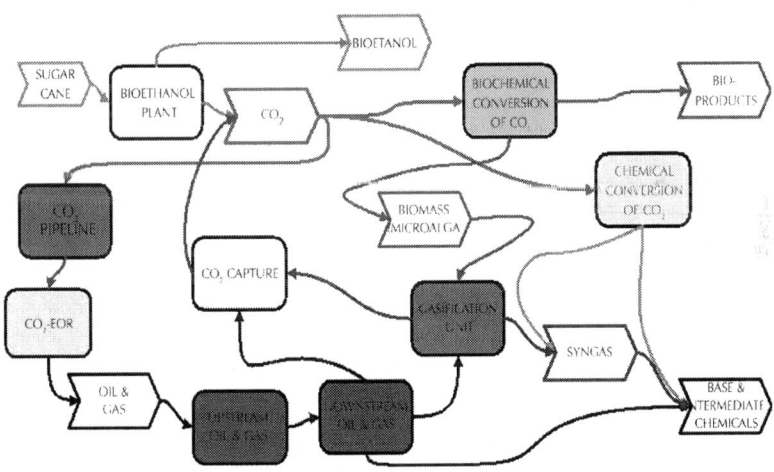

Figure 3: CO_2 capture and utilization in the oil & gas supply chain.

EOR and CGS constitute, under present conditions, the only CO_2 processing chain that has some steps ready to be put into operation at high scales. These encompass some separation technologies for CO_2 post-combustion capture, CO_2 compression and CO_2 transportation via long pipelines. EOR and CGS can be reinforced if oxy-fuel technologies finally attain industrial maturity, which may occur within the next 5 decades. Pre-combustion technologies are also potential powerful contributors to EOR-CGS systems, which should be fully developed within the next 5 decades. Pre-combustion relies on some separation technologies also present in post-combustion alternatives.

BC via solar photosynthesis is a promising but still incipient package of technologies that have yet to be proven feasible in large scales. The main drawback is the impressive footprint and liquid hold-up of algae cultivation and processing plants, which have to comply with the high dispersion and periodicity of solar light and small concentration of biomass in growing medium (always below 10g/L). Besides, the use of non-solar light energy is totally out of question due to the photosynthesis efficiency limitations of green life forms (always below 10%, i.e., only a maximum of 10% of the incident light, already with appropriate wavelength, is biochemically usable to bio-convert CO_2).

CC context encompasses novel approaches extending the technology scenario for CO_2 supply chain via chemical conversion to benign, stable compounds for long-term storage or to value-added products like plastics, chemical intermediates and oxygenated octane enhancers like DMC – dimethyl-carbonate (Souza et al., 2013) for reuse.

The cPSE vision of CO_2 treats routes of CC and BC (and BC integrated with CC, i.e., BCC) to useable products and fuels, not as substitute, but rather as a complement to EOR and CGS. The present text is oriented to analyze CO_2 as a profitable feedstock, i.e., its potential use as feedstock to chemicals and fuels within economic applications, besides its relevant use in EOR. CGS is not considered an economic application; instead, it constitutes only a plausible, secure, destination of the excess of carbon, i.e., all CO_2 that has to be discarded because it is not dispatched to any economic use.

Among the routes for CO_2 utilization, BC is a natural choice as photosynthesis yields biomass, allowing the production of bioproducts, biofuels and chemicals through downstream CC processing routes. In fact, biomass gasification is the most flexible technology for dropping into conventional downstream CC routes. This integrated option configures BCC – biochemical and chemical conversion of CO_2.

The beneficial use of alkaline wastes or metallic ions to convert CO_2 via neutralization of alkaline wastes, or reaction of CO_2 with metallic ions to form less soluble carbonates that can be removed from produced water (oil & gas industry) is also a relevant CC application of CO_2. Lastly, this study presents the EOR use of CO_2, i.e., by injection into depleting oil or gas fields to maximize hydrocarbon production.

Finally, analysis of routes that undertake CO_2 reduction must take into account the life cycle of the processes in order to assess whether additional CO_2 production occurs beyond the amount abated from atmospheric emissions. This is precisely the case of CC of CO_2 into fuels and chemicals that always requires high energy input, normally derived from burning fossil fuels, entailing further associated GHG emissions. In this context, new or mature cPSE solutions should comply with the triple objective of sustainability, namely: economically feasible, environmentally benign and socially beneficial, in a supply chain approach. In connection with this, the chapter presents the CO_2 Capture Cycle as well as promising alternatives of its reutilization.

THE THERMODYNAMICS OF PURE CO_2

The phase behavior of pure CO_2 (Figure 4) exhibits particularities when compared with common light species of natural gas (NG) like CH_4, C_2H_6 and C_3H_8. In general grounds, CO_2 has a fluid phase behavior similar to ethane (C_2H_6) with a very similar Critical Point (CP) temperature. The pronounced distinctive characteristic is its high Triple Point (TP) (Figure 4) temperature comparatively with light hydrocarbons. This means that solid CO_2 (dry ice) can be easily encountered below -56.6°C and above 5.2bar, if the original processing stream is rich enough in CO_2. This freeze-out of solid is not observed with the light hydrocarbon species unless below -182°C, which is about 20°C below the lowest temperature that can be achieved in LNG processing (i.e., the Normal Boiling Point of CH_4). In other words, without CO_2, the coldest NG processing (LNG plant) does not have solid formation. The phase behavior of pure CO_2 (Figure 4) is characterized by two larger (and infinite) continents corresponding to gas and solid states, and a finite intermediate liquid continent extending between the TP and CP temperatures. The three continents are two-dimensional (2D) objects due to the Phase Rule, which stipulates two degrees of freedom for a pure species at one-phase condition. One-dimensional (1D) equilibrium boundaries – SLE, SVE and VLE lines – are positioned between two neighboring continents. Their 1D nature is also a consequence of the phase Rule, which stipulates one degree of freedom for a pure species at a two-phase condition. SLE – solid-liquid (or solid-fluid) equilibrium line – is an endless line extending from the TP to indefinitely high pressures, characterized by the coexistence of solid and liquid CO_2. SVE – solid-vapor equilibrium line – is characterized by coexistence of solid and gas CO_2 ending at the TP and lower bounded by the absolute 0K. VLE – vapor-liquid equilibrium line – is a finite locus between TP and CP where liquid and gas CO_2 coexist. Just above the liquid continent there is a somewhat indefinite supercritical fluid zone (SCF). The SCF is a fusion of gas and liquid behaviors at high pressures above the CP pressure and high temperatures above the CP temperature. The SCF is characterized by high densities and high compressibility.

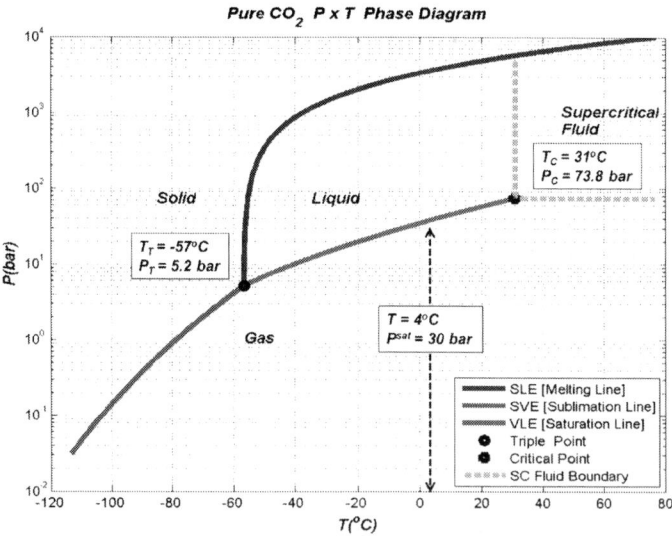

Figure 4: Phase behavior of pure CO_2 [TC, PC critical point; TT, PT triple point].

The contact point for the three continents is the Triple Point TP, an invariant three-phase point (zero degrees of freedom) by the Phase

Rule. The VLE locus ends at the Critical Point CP where the differences between liquid and vapor vanish. The CP is also an invariant point with zero degrees of freedom, but, contrary to the TP, the CP is a single phase point satisfying two extra criticality conditions. The phase behavior of CO_2 depicted in Figure 4 can be enriched if density (kg/m^3) is put on the third axis as shown in Figure 5. In Figure 5, densities of gas phase are calculated via the Peng-Robinson (PR) equation of state (EOS), whereas liquid and solid phase densities (saturated or not) are calculated via the correlations presented by Span and Wagner (1996) and Trusler (2011), respectively. In Figure 5, the saturation lines SLE, SVE and VLE are also depicted, where the VLE line is presented with its gas and liquid branches merging at the CP. Figure 5 reveals that CO_2 can be found with densities well above the density of water (1000 kg/m^3), either as a solid or as high pressure dense liquid or SCF fluid.

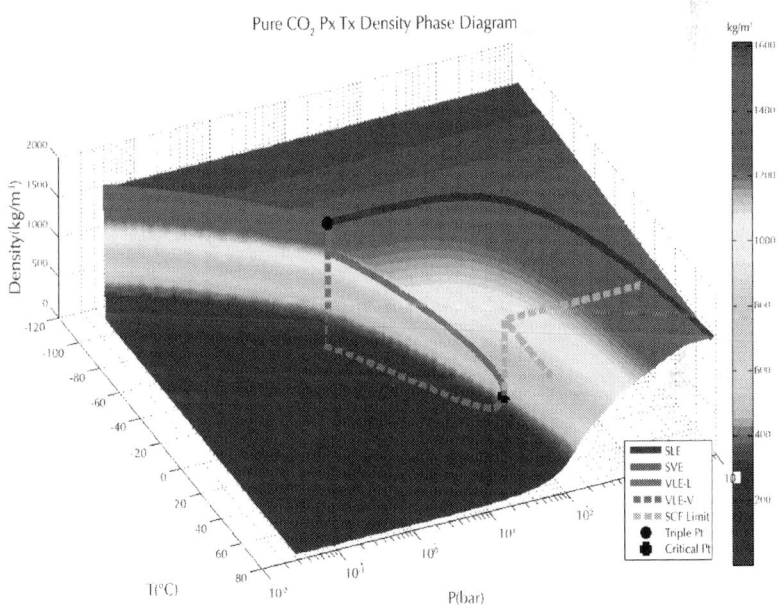

Figure 5: Density x Pressure x Temperature of Pure CO_2 [with TP and CP points, VLE, SVE & SLE lines and SCF domain].

The PR-EOS is a very simple thermodynamic relationship that can be used to predict gas and liquid properties of pure CO_2 and of CO_2 rich mixtures. The PR-EOS is certainly not the most accurate state relationship to address the fluid properties of CO_2 (Genesis, 2011; see also the EOS presented by Span and Wagner, 1996), but it does represent the best compromise between simplicity of use and accuracy for CO_2 rich systems without water, either in single fluid phase or in two coexisting fluid phases (VLE), at low or high pressures and densities (Li, 2008; Li and Yan, 2009; Genesis, 2011). According toGenesis (2011), the PR-EOS produces errors for compressibility factor (Z), molar enthalpy (H) and sound speed (C) below 1% for CO_2 subcritical gas phases and between 5% and 20% for subcritical liquid phases. Near the CP, the errors in Z, H, C can be higher but the uncertainties of experimental values are also higher. In the SCF domain, the PR errors for Z, H, C fall between 1% and 15%. In all the aforementioned cases, errors are relative to Span-Wagner EOS for fluid CO_2. The PR-EOS also seems adequate to describe the supercritical fluid (SCF) domain of CO_2 and its rich mixtures near critical transitions. This means that PR-EOS can be used to address any property (density, enthalpy, entropy, exergy, etc) of liquid and vapor phases of CO_2 and its mixtures with NG species under moderate errors, which are compatible with engineering applications (Li, 2008 and Li and Yan, 2009;Genesis, 2011). The PR-EOS is presented in Eq. (1), while its classical mixing rules follow in Eqs. (2)and (3). PR-EOS component parameters are given in Eq. (4) to (6) where *Tci*, *Pci*, *I* and *nc* are, respectively, critical temperature, critical pressure, acentric factor for component *i*, and the number of components. *Kij* represents the binary interaction parameter (BIP) for species *i* and *j* which is symmetric, can be used as zero in some cases, and is not necessary for pure CO_2. In Eqs. (1) to (3), *V, T,P,R* and *Ni* represent volume (m³), temperature (K), pressure (bar), ideal gas constant ($8.314.10^{-5}$bar.m³/mol.K) and the mol number of species *i*.

$$P = \frac{NRT}{V - Nb} - \frac{N^2 a}{V^2 + UVNb + W(Nb)^2}, \quad U = 2, \quad W = -1 \tag{1}$$

$$Nb = \sum_{i=1}^{nc} N_i b_i \tag{2}$$

$$N^2 a = \sum_{i}^{nc}\sum_{j}^{nc} N_i N_j \sqrt{a_i}\sqrt{a_j}\sqrt{\Phi_i(T)}\sqrt{\Phi_j(T)}\left(1 - K_{ij}\right), \quad (K_{ij} = K_{ji}) \tag{3}$$

$$a_i = 0.45724 \frac{R^2 Tc_i^2}{Pc_i}, \quad b_i = 0.07780 \frac{R Tc_i}{Pc_i} \tag{4}$$

$$\Phi_i(T) = \left(1 + g(\omega_i)\left(1 - \sqrt{T/Tc_i}\right)\right)^2 \tag{5}$$

$$g(\omega_i) = 0.37464 + 1.54226\omega_i - 0.26992\omega_i^2 \tag{6}$$

If dimensionless terms in Eq. (7) are used in Eq. (1), the classic Z cubic form results in Eq. (8):

$$Z = \frac{PV}{NRT}, \quad B = \frac{PNb}{NRT}, \quad A = \frac{PN^2a}{(NRT)^2} \tag{7}$$

$$Z^3 - (1 + B - UB)Z^2 + (A + WB^2 - UB - UB^2)Z - AB - WB^2 - WB^3 = 0 \tag{8}$$

THE CO_2 CAPTURE CYCLE

According to Oi (2010), CO_2 removal from process streams at an industrial scale has occurred since about 1930, mainly from natural gas and from industrial gases at high pressures for ammonia and methanol production. Like any separation unit in any kind of process, the CO_2 capture technology has to be judiciously chosen as it may undermine the profitability, controllability, safety, and simplicity of the plant. Nevertheless, even when properly selected, separations usually raise concerns like heat and mechanical energy consumption, increased utility use, carbon emission, chemicals demands, size, weight, footprint, construction restraints, operational hazards, etc.

The CO_2 capture cycle encompasses two or three main unit operations, which have to separate CO_2 from the gas mixture and send it to an appropriate destination. First, there is the CO_2 transfer across the gas phase into the medium that contains the binding material: a solvent or an adsorbent or a selective barrier. Second, there is (or not) the regeneration of the binding medium with concomitant CO_2 release. Third, there is the compression and cooling of the captured CO_2 because it has to be handled at high density or as a liquid.

Technologies for CO_2 capture from gas streams include chemical absorption (e.g., aqueous ethanolamines and aqueous K_2CO_3), physical absorption (e.g., propylene carbonate, selexol and rectisol), physical adsorption, membrane permeators, membrane contactors, cryogenic distillation and hybrid technologies (e.g., membrane permeator followed by ethanolamine absorption). Among those, the most relevant technologies include membrane permeation and chemical absorption with aqueous ethanolamines, the later standing as the most mature CO_2 capture technology from gas streams.

Chemical Absorption with Aqueous Alkanolamines

With regards to the solvent, the advantages of the chemical absorption of acid gases – CO_2 and H_2S – by aqueous alkanolamines are well-known: the former are weak acids and the later are weak alkali, such that they reversibly bind at low temperature and high acid gas fugacity and subsequently untie at higher temperatures and low fugacity, leading to efficient acid gas stripping, which regenerates the solvent. The relevant variable in the liquid phase is the solvent loading (in mol of acid gas per mol of amine), which expresses the degree of conversion of amine in the solvent. Typically, the loading assumes values in the range 0 to 1.2 mol/mol amine.

AGWA (Acid Gas, Water and Amines) systems is a convenient denomination (de Medeiros et al., 2013a; de Medeiros et al., 2013b) of such reactive vapor-liquid equilibrium (RVLE) systems containing Acid Gas, Water and Amines. Amines are understood to be the common alkanolamines like monoethanolamine (MEA, MW=61), diethanolamine (DEA, MW=105), methyl-diethanolamine (MDEA, MW=119) and 1-amino-2-propanol (AMP, MW=75).

MEA is a benchmark co-solvent for CO_2 capture, with: (i) satisfactory absorption capacity; (ii) fast kinetics; (iii) miscible with water in all proportions, and (iv) low cost. On the other hand, MEA is problematic in terms of solvent regeneration, because it exhibits: (i) the highest energy load per unit of stripped gas; (ii) corrosion and chemical/thermal degradation concerns; (iii) non-negligible evaporation losses due to its low boiling point; and (iv) high reactivity entailing low H_2S/CO_2 selectivity. As a secondary amine, DEA is less reactive than MEA but it has the following comparative advantages: (i) lower energy per unit of stripped gas; (ii) more resistant to degradation; (iii) less corrosive; and (iv) less volatile. MDEA exhibits the lowest reactivity with CO_2 and the greatest resistance to degradation. When compared to DEA and MEA, MDEA presents the following advantages: (i) the highest equilibrium capacity of acid gas absorption (in the case of CO_2, nearly two times the capacity of primary amine); (ii) lowest regeneration cost (does not form stable products with CO_2); (iii) high H_2S/CO_2 selectivity thanks to its low reactivity with CO_2; (iv) lowest enthalpies of reaction and lowest regeneration heat per unit of stripped gas; and (v) negligible losses due to very low vapor pressure. Alkanolamines with steric hindrance like AMP show reduced carbamate stability and the methyl adjacent to the amine group may affect absorption capacity and/or its rate. AMP exhibits absorption capacity and stripping heat similar to MDEA, but a faster reaction with CO_2 during the capture step (Medeiros et al., 2013b).

AGWA chemical reactions are really three-reactant transformations in the liquid phase with 1:1 amine to acid gas mol ratio. Water is necessary at high mol ratio to amine (8:1 or 7:1), otherwise the reaction simply does not evolve sufficiently, leaving non-solvated amine unconverted (low loading). This is the reason why MEA concentration in the solvent is upper bounded at 30%w/w in water (or 11.2%mol MEA + 88.8%mol H_2O), and 50%w/w in the case of MDEA (or 13.1%mol MDEA + 86.9%mol H_2O). Figure 6 presents a schematic of the chemical absorption system for CO_2/H_2S capture. AGWA absorption reactions take place in the colder higher pressure absorption column, whereas in the hotter lower pressure stripper column, AGWA reactions are reverted, thus releasing free acid gas and regenerating amine at the bottom. The main consumption of heat occurs in the reboiler in the bottom of the stripper column, where water is vaporized breaking the liquid phase association of acid gas, water and amine. Consequently, typical heat consumptions of MEA strippers lay between 167kJ and 200kJ per mol of stripped CO_2 (or between 3.8 GJ and 4.5 GJ/tonne of stripped CO_2). These figures are impressive high values, equaling 4 to 5 times the molar heat of vaporization of water per mol of stripped CO_2.

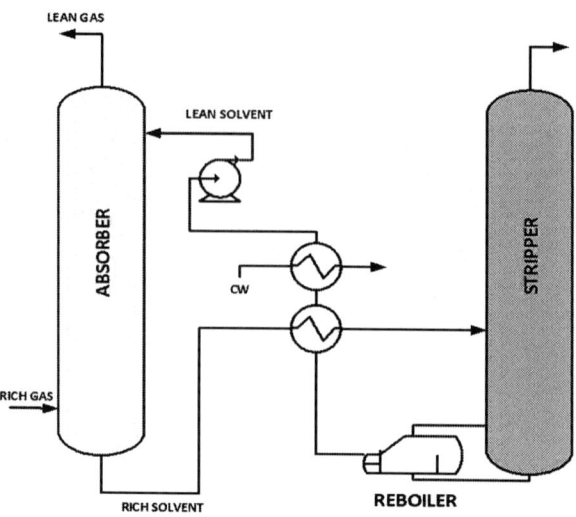

Figure 6: CO_2 capture by chemical absorption.

The literature presents several modeling approaches for absorption and stripping with AGWA systems. The most common approach involves cumulatively ionic species within ideal solution, ideal gas vapor phase, reversible chemical kinetics and rate-based interfacial mass transfer (de Medeiros et al., 2013b). This kind of approach is classical and is more adequate to low pressure and to dilute AGWA systems as in CO_2 capture from combustion gases. For high capacity and high pressure AGWA with rich CO_2 natural gas systems, high loadings and high heat effects may appear. For such AGWA systems de Medeiros et al. (2013a, 2013b) proposed a molecular Chemical Theory approach where molecular complex species are formed in the liquid phase via chemical equilibrium reactions having as reactants real AGWA species CO_2, H_2S, H_2O, MEA, DEA, MDEA and AMP. Each complex species is created by reacting three real species: an acid gas, water and an amine, as shown in Eq. (9). These complexes are reversibly created during the absorption step and destroyed during the stripping step. The main advantages of the molecular chemical theory of de Medeiros et al. (2013b) are:

- Theory is a Reactive Vapor-Liquid Equilibrium (RVLE) framework that employs only VLE AGWA data for model tuning via implicit statistical procedures. Model can be tuned with low or high pressure data, appropriate either for the absorption column (low

temperature, high pressure) or for the stripping column (high temperature, low pressure).
- AGWA RVLE is, in fact, an ionic scenario dominated by weak ions and several unknown solvated complexes. Thus, the limited knowledge to handle all the possible ions is circumvented via such nonionic, nonvolatile complexes.
- Model tuning requires only AGWA VLE data, much more available than non-equilibrium counterparts.
- The nonionic assumption is proposed to mimic the weakness of AGWA electrolytes, all created by incomplete dissociations.
- All thermodynamics properties (of both vapor and liquid phases) in the formalism of de Medeiros et al. (2013b) are calculated via cubic EOS (Soave-Redlich-Kwong - SRK and Peng-Robinson - PR), because all species considered are purely molecular.
- The formalism can be used with both acid gases CO_2 and H_2S and with blends of alkanolamines, which are used for conjugating desirable qualities like good reactivity of MEA with lower costs of regeneration and corrosion of MDEA. Such blends are promising in terms of gains relatively to individual amines.

The AGWA formalism of de Medeiros et al. (2013b) was calibrated with VLE AGWA data from Literature. These data configure a large database of AGWA-VLE with 1331 runs shown in Figure 7. This database includes several runs with alkanolamines (MEA, DEA, MDEA and AMP), water and two acid gases or solutes (CO_2 and H_2S) at pressures ranging from 0.1 bar to 30 bar and temperatures from 25°C to 140°C. The calibration parameters correspond to chemical equilibrium constants belonging to chemical reactions that convert acid gas, water and amine into molecular complex species in liquid phase as shown in Eqs. (9). These chemical reactions are chemical equilibrium (ChE) reactions, which evolve to the right when CO_2 and H_2S are absorbed by the solvent and to the left when CO_2 and H_2S are stripped from the solvent by the action of heat and low pressure. In other words, the set of chemical equations in Eq. (9) can reproduce both phenomena absorption and solvent regeneration. This is a physically sound approach that allows, among other things, to estimate heat effects that occur in these processes like the release of heat during gas absorption and the absorption of heat during stripping of CO_2 and H_2S.

$$CO_2(g) + H_2O(g) + MEA(g) \xleftrightarrow{K_1} CO_2 - H_2O - MEA(Liq)$$
$$CO_2(g) + H_2O(g) + MDEA(g) \xleftrightarrow{K_2} CO_2 - H_2O - MDEA(Liq)$$
$$CO_2(g) + H_2O(g) + DEA(g) \xleftrightarrow{K_3} CO_2 - H_2O - DEA(Liq)$$
$$CO_2(g) + H_2O(g) + AMP(g) \xleftrightarrow{K_4} CO_2 - H_2O - AMP(Liq)$$
$$H_2S(g) + H_2O(g) + MEA(g) \xleftrightarrow{K_5} H_2S - H_2O - MEA(Liq) \quad (9)$$
$$H_2S(g) + H_2O(g) + MDEA(g) \xleftrightarrow{K_6} H_2S - H_2O - MDEA(Liq)$$
$$H_2S(g) + H_2O(g) + DEA(g) \xleftrightarrow{K_7} H_2S - H_2O - DEA(Liq)$$
$$H_2S(g) + H_2O(g) + AMP(g) \xleftrightarrow{K_8} H_2S - H_2O - AMP(Liq)$$

The parameter estimation of the AGWA model of de Medeiros et al. (2013b) involves estimating ChE constants for the chemical reactions in Eq. (9) at selected temperatures. Thus, several experimental AGWA VLE points at a chosen temperature are extracted from the AGWA database in Figure 7 and are processed via a maximum likelihood algorithm to estimate the ChE constants. The estimation algorithm also assures that the set of nonlinear constraints of each selected experiment is satisfied. The set of constraints of a given experiment is shown in Table 1 with Eqs. (10) to (16) including: (i) Real species – CO_2, H_2S, H_2O, MEA, DEA, MDEA, AMP – mass balance (RMB); (ii) mol fraction normalization for each phase (SXY); (iii) definition partial pressures of solutes (acid gases) (PPS); (iv) definition of loadings of solutes (LDG); (v) VLE of Real species (VLE); (vi) ChE of Complex formation in Eq. (9) (CHE). The nomenclature used in Table 1 is shown in Table 2, including the matrix of stoichiometric coefficients of Eq. (9) in Eq. (10).

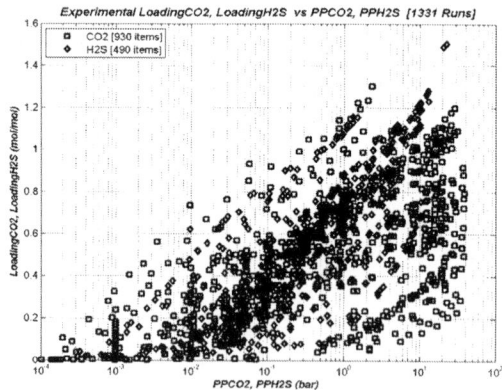

Figure 7: AGWA Database used to estimate parameters in the Chemical Theory Model of de Medeiros et al. (2013b) [PPCO2, PPH2S – partial pressures (bar) of CO_2 and H_2S].

Figures 8 and 9 depict some results after calibration of the AGWA model of de Medeiros et al. (2013b) with the database of AGWA experiments in Figure 7. Both figures refer to parameter estimation using 320 experimental AGWA data points at 40°C. Figure 8 presents experimental versus predicted loadings (mol/mol amine) of solutes CO_2 and H_2S at 40°C, whereas Figure 9 presents experimental versus predicted partial pressures (bar) of solutes CO_2 and H_2S at 40°C. As can be observed, there is a satisfactory agreement between experimental values and predicted counterparts.

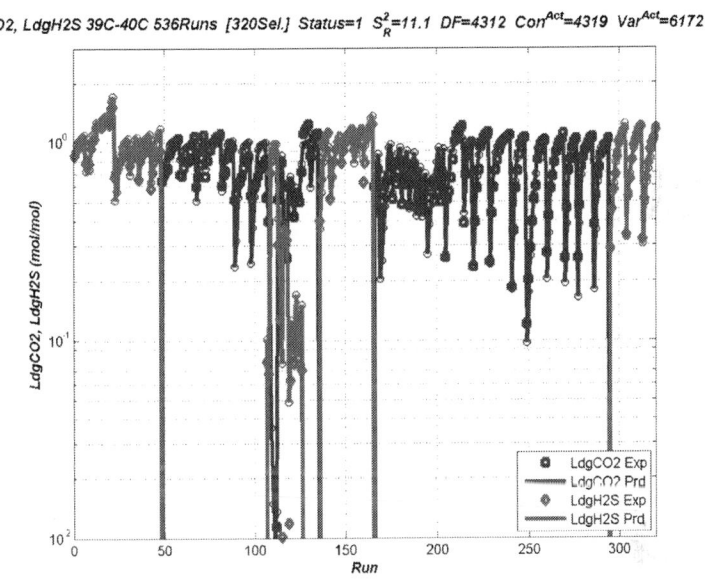

Figure 8: Experimental vs predicted Loadings of CO_2 (blue) and H_2S (red) after calibrating the AGWA model at 40°C with 320 data points (de Medeiros et al., 2013b) [LdgCO2, LdgH2S – Loadings of CO_2 and H_2S (mol/mol amine)].

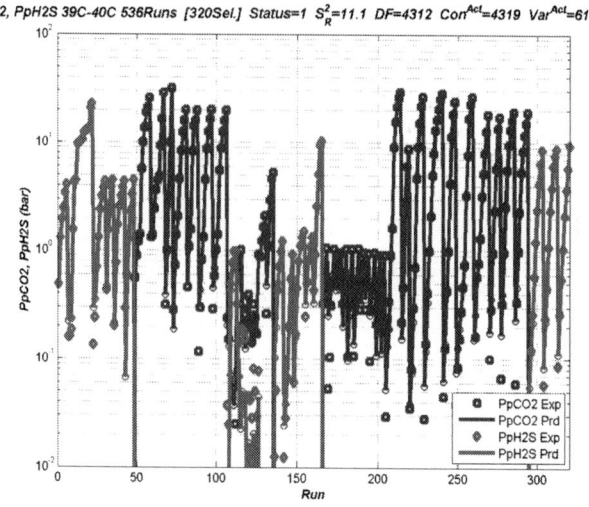

Figure 9: Experimental vs predicted partial pressures of CO_2 (blue) and H_2S (red) after calibrating the AGWA model at 40°C with 320 data points (de Medeiros et al., 2013b) [PpCO2, PpH2S – Partial pressures of CO_2 and H_2S (bar)].

Predicted values are calculated solving the set of constraints of the AGWA model in Eqs. (I) to (VIII) with all thermodynamic properties (e.g. vapor and liquid fugacities) estimated with SRK-EOS. The estimated parameters are the eight ChE constants in Eq. (9) and (VIII) at the corresponding temperature of 40°C.

Table 1: Set of Constraints of AGWA VLE Experiments [real species ($n=7$): CO_2, H_2S, H_2O, MEA, DEA, MDEA, AMP], [solutes ($ns=2$): CO_2, H_2S], [Complexes ($nr=8$): see Eq. (9)]

Constraints [Label]	Expression	Eq.
n Real Species Balances [RMB]	$\underline{N} + \underline{\underline{L}}\,\underline{\underline{\Pi}}\,\underline{X}_C - \underline{\underline{L}}\,\underline{X} - \underline{\underline{V}}\,\underline{Y} = \underline{0}$	I
	$\sum_{j=1}^{n} X_j + \sum_{k=1}^{n_C} X_{Ck} - 1 = 0$	II

Normalizations [SXY]	$\sum_{j=1}^{n} Y_j - 1 = 0$	III
n_s Solute Partial Pressures [PPS]	$\underline{P}_S - P\underline{\underline{S}}_S \underline{Y} = \underline{0}$	V
n_s Solute Loadings [LDG]	$(\underline{1}^T \underline{\underline{S}}_A \underline{N}) \underline{\alpha}_S - \underline{\underline{S}}_S (\underline{N} - V\underline{Y}) = \underline{0}$	VI
n VLE of Real Species [VLE]	$\ln \hat{\underline{f}}^V - \ln \hat{\underline{f}}^L = \underline{0}$	VII
n_r Reactions ChE [CHE]	$\ln \underline{X}_C + \underline{\underline{\Pi}}^T \ln \hat{\underline{f}}^L - \ln \underline{K}(T) = \underline{0}$	VIII

Table 2: Nomenclature used in Table 1

Symbol	Definition
$\underline{\alpha}_S$	Vector of solute loadings (mol/mol of amine)
$\hat{\underline{f}}^V, \hat{\underline{f}}^L$	Vectors of Real species fugacities in vapor and liquid phases (bar)
$\underline{K}(T)$	Vector of ChE constants of Eq. (9)
L	Liquid phase mole number
n_S, n, n_r	Numbers of solutes (=2), Real species (=7) and Complex species (=8)
$\underline{N}$	Vector of total mole number of Real species in the experiment
$\underline{P}_S$	Vector of solute partial pressures (bar)
P	Pressure (bar)

$\underline{\underline{\Pi}}$	Matrix of stoichiometric coefficients of Eq. (9), shown in Eq. (17).
$\underline{\underline{S}}_A , \underline{\underline{S}}_S$	Selection matrices for amines and solutes with sizes 4 X 7 and 2 X 7
V	Vapor phase mole number
$\underline{X}, \underline{Y}$	Vectors of mol fractions of Real species in liquid and vapor phases
$\underline{X}_C$	Vector of mol fractions of Complex species in liquid phase

$$\underline{\underline{\Pi}} = \begin{bmatrix} -1 & -1 & -1 & -1 & 0 & 0 & 0 & 0 \\ 0 & 0 & 0 & 0 & -1 & -1 & -1 & -1 \\ -1 & -1 & -1 & -1 & -1 & -1 & -1 & -1 \\ -1 & 0 & 0 & 0 & -1 & 0 & 0 & 0 \\ 0 & -1 & 0 & 0 & 0 & -1 & 0 & 0 \\ 0 & 0 & -1 & 0 & 0 & 0 & -1 & 0 \\ 0 & 0 & 0 & -1 & 0 & 0 & 0 & -1 \end{bmatrix} \qquad (10)$$

The huge heat consumption of MEA strippers (167 kJ to 200kJ/mol of stripped CO_2) forces investigation on more efficient alternatives of solvent regeneration. Wagener and Rochelle (2010) recognize as a "monumental task" reducing energy penalty of CO_2 capture from coal-fired power plants (approximately 30%). They presented an analysis of various stripper configurations, concluding that increasing complexity improves performance at the cost of higher capital and operational expenditures, i.e., an optimal scheme should exist. The alternative configurations include simple stripping with vapor recompression, multi-pressure, double matrix, split product, internal exchange, and flashing feed. Wagener and Rochelle (2010) concluded that operating with multiple pressure levels reduces the energy requirement as "equivalent work" (including reboiler duty, pumping and heat exchangers) in 33.6 kJ/molCO_2 captured, with an optimal lean loading of 0.375 mol/mol amine. Moreover, they claim that the arrangement benefits from stripping at high pressure, whilst improves reversibility when returning to atmospheric conditions for the absorber. Wagner and Rochelle (2011) revisited several configurations with varying levels of complexity and reported that an inter-heated column and a simple stripper required 33.4 kJ/mol CO_2 and 35.0 kJ/mol CO_2 of equivalent work, respectively, at their optimum lean loadings.

Membrane Permeators

The first membrane modules were developed as planar films. However, such arrangement has a low ratio of membrane transfer area per equipment volume. Presently, the majority of modules for CO_2 gas separation are manufactured in hollow-fiber or spiral wound configurations. Figure 10 shows a schematic of a hollow-fiber membrane permeator module and the selective transport of molecules across the hollow-fiber membrane.

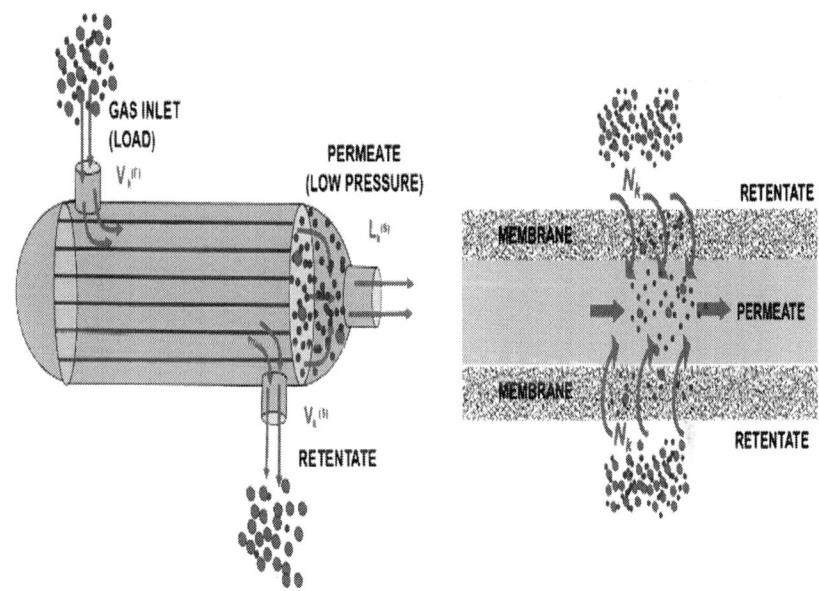

Figure 10: Hollow-fiber membrane permeator and the transfer through a hollow-fiber.

The membrane acts selectively against gas diffusion from the high pressure side (the retentate) to the low pressure side (the permeate) creating separation. The membrane – either in hollow-fiber or spiral wound configurations – is composed by two polymeric structures: a permselective dense (nonporous) skin over a thicker microporous substrate. Inside the permeator gas coming from the high pressure retentate stream solubilizes into the dense skin and diffuses to the low pressure side creating an almost adiabatic expansion with consequent fall of temperature via Joule-Thomson effect. Such fall of temperature may create problems to the proper functioning of the module due to eventual condensation of less volatile species that are retained and accumulate in the retentate. This condensation is undesirable and can damage the membrane plastic material.

Ho and Wiley (2005) evaluated the economic performance of membrane separation for low pressure (flue gas applications) and high pressure (e.g., natural gas processing). The authors concluded that the highest share in capital cost was due to the compression phase (around 80%) while membrane and the respective shells exhibited 10% of the investment. For natural gas processing, however, membrane has the largest share of CAPEX (about 62%), as the compression is an existing stage of natural gas processing. Moreover, Ho and Wiley also report an alternative multi-stage configuration in order to obtain a CO_2 rich permeate in a second membrane stage, which demands a recompression step corresponding to 30% of the total CAPEX.

In oil and natural gas deep-water offshore rigs, an issue in the processing of large volumes of NG with high CO_2/CH_4 ratio vis-à-vis climate concerns, involves large capacity CO_2 separators, whose targets are exportation of saleable NG from the retentate product via long pipelines and feeding enhanced oil recovery (EOR) systems with hyperpressurized CO_2 from the permeate product. In this scenario, robustness, angular indifference, modularity, and compactness also influence the selection of a separation technology. In this context, membrane permeation batteries are usually favored against the more traditional absorption columns with amine solvents (Medeiros et al., 2013a). On the other hand, large permeation batteries also have their own shortcomings, mostly related to the permselective dense (nonporous) skin over the microporous substrate. The permselective skin forces the existence of a high P and exhibits low fluxes which means low capacity per unit area and high consumption of power for permeate recompression and/or permeate recycles. It also has CO_2/CH_4 limited selectivity and demands feed necessarily dew-point conditioned (to prevent gas condensation in the retentate provoked by the decrease of temperature associated with permeation) and continuous inspection looking for membrane bursts.

A model for hollow-fiber permeators can be built by writing mass, momentum and energy conservation principia for both permeate and retentate. Additionally, an appropriate model for trans-membrane flux transport is also necessary. Typically, trans-membrane flux models are written as products of a permeancy coefficient and a driving force term. Each species has a permeancy coefficient for a given membrane and given conditions of temperature, pressure and gas composition, but, in general, permeances are assumed independent of gas composition and pressure. The driving force term is usually expressed as a difference of component fugacities between the retentate and permeate sides. In this context and assuming a co-current compressible flow of both permeate and retentate, Nakao et al. (2009) proposed a stationary hollow-fiber membrane permeator rigorous model which can be used to predict CO_2 separation from CO_2 rich NG. The model of Nakao et al. (2009) builds a spatial 1-dimensional axial description where species mass balances, energy and momentum balances are rigorously written for both permeate and retentate. All permeate and retentate thermodynamic properties (enthalpies, densities, fugacities, etc) are rigorously calculated via PR-EOS, Eq. (1). The model is too involved to be discussed here in detail, but the clarity and usefulness of it results are worth presenting in the context of CO_2 capture. Figure 11 depicts a natural gas feed named GBS2 (from Basin of Santos, Brazil) with 2.79554 MMNm³/d at 53bar and 38°C with, initially, 25.32%mol CO_2 and 55.64%mol CH_4. Figure 11 depicts the VLE locus of GBS2 and the heat capacity (Cp in kJ/mol.K) map versus $T \times P$ with vivid identification of the critical and supercritical neighborhoods due to the second-order transition across the critical state that is revealed by second order properties like Cp. In Figure 11 [A], the gas feed is located at 28°C above its dew boundary (blue). GBS2 passes through a battery of 40 horizontal modules (0.2m X 10m each) of cellulose-acetate hollow-fibers (0.5mm ID), with CO_2 and CH_4 permeances respectively of $1.27.10^{-8}$ and $4.4.10^{-10}$ mol/s.m².Pa and 1854m² of permeation area per module. Permeate pressure is 4bar and the external temperature is 27°C with a heat transfer coefficient of 5W/m².K. Resulting profiles in the axial flow direction (10m long) for one module are shown in Figure 12. Permeate and retentate initiate contact at axial position $X=0m$ and cease contact at $X=10m$. The background along the axis of flow is painted in four colors for discrimination among the four quarters of a typical module. Figure 12 [A] depicts retentate profiles of %mol CO_2

and CH_4 showing a CO_2 decrease from 25.62% to 9.3%mol and a CH_4 increase from 55.64% to 65.5%mol. Figure 12 [B] shows that the final permeate recoveries of CO_2 and CH_4 are 72% and 11.5%. Figure 12[C] depicts permeate and retentate temperature profiles revealing a drop of retentate temperature from 38°C to 29°C, insufficient for condensation (Figure 11 [A]).

Gas-Liquid Contactors

Gas-liquid contactors (GLC) constitute a new and versatile kind of membrane operation for CO_2 removal from NG. A GLC unit admits a gas stream which is contacted with a liquid stream (the solvent stream) through a polymeric micro-porous membrane. The solvent phase is an aqueous solution of ethanolamines (e.g. MEA and/or MDEA) that can absorb CO_2 as occurs in an absorption column. But the difference here is that the liquid and gas phases really do not mix as they do in direct contact devices like packing towers. Assuming that the GLC is manufactured with hollow-fiber membranes, the solvent phase flows in the hollow-fiber inner space while the gas phase flows in the outside shell space, but there are also configurations where the roles of liquid and gas phases are inverted. A module of hollow-fiber GLC with gas flowing in the shell side and with solvent flowing in the inner space of the hollow-fibers is sketched in Figure 13. As can be seen, the module resembles closely a shell and tube indirect contact heat exchanger.

GLC membrane operation can outperform common nonporous permeators in terms of capacity per unit area while sustaining high CO_2/CH_4 selectivity. High fluxes are possible in GLC because the membrane does not have a dense skin in order to be selective. The underlying reason is that selectivity is imposed by the solvent in the inner (permeate) side cutting the necessity of high P across the membrane. As the reader can see, terms permeate and retentate are also used here despite some improperness, and refer, respectively, to the inner solvent flow and to the gas flow that was not transferred to the inner solvent.

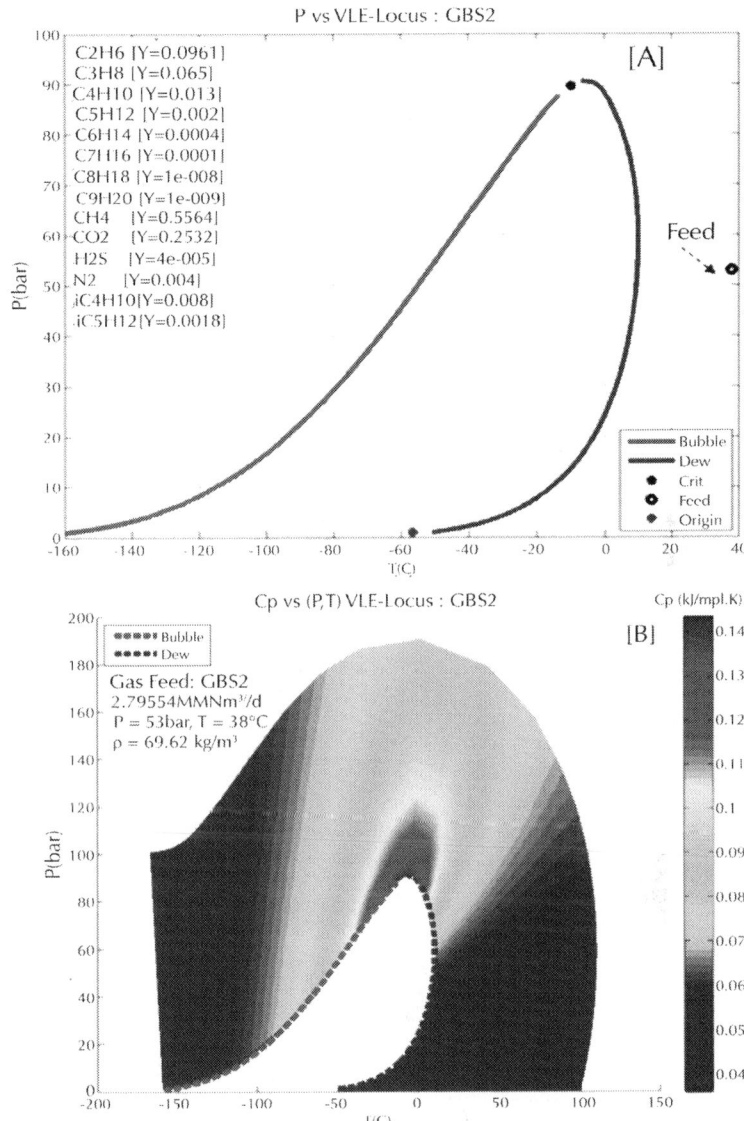

Figure 11: GBS2 a CO_2 rich natural gas (25.32%CO_2, 55.64%CH_4, 9.61%C_2H_6, 9.43%C_3^+) for Hollow-Fiber Permeator: [A] VLE Locus P(bar) X T(°C); [B] CP(kJ/mol.K) vs P(bar) X T(°C).

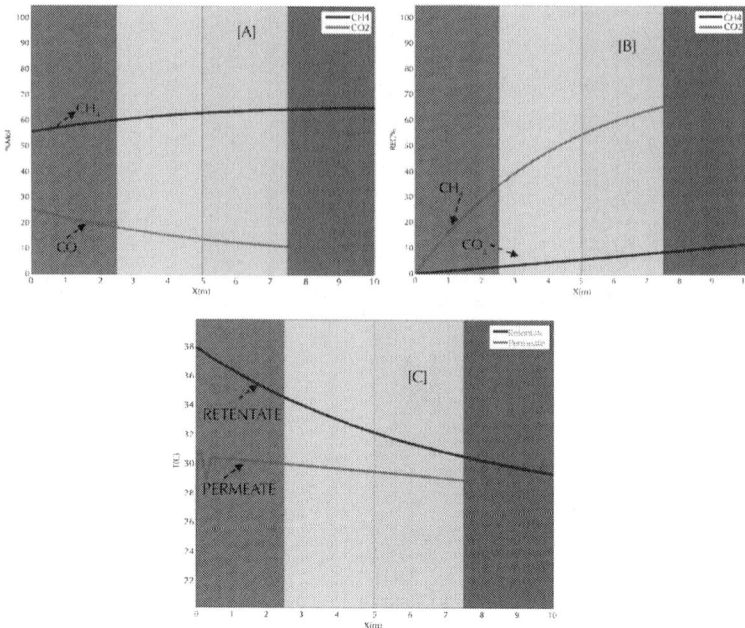

Figure 12: Profiles in a hollow-fiber module in co-current flow with feed GBS2 in Fig. 3.6 (25.32% CO_2 & 55.64% CH_4): [A] CO_2 & CH_4 %mol in retentate; [B] CO_2 & CH_4 permeate % recoveries; [C] retentate and permeate temperatures.

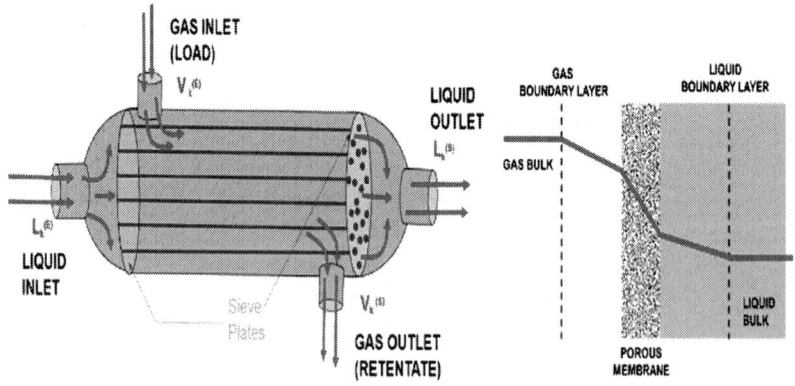

Figure 13: Hollow-fiber gas-liquid contactor with solvent in the inner space of membrane.

GLC can separate CO_2 from NG offering the advantages of both membrane and absorption technologies, but leaving behind the respective shortcomings like the low flux coexisting with high difference of pressure between the membrane sides and the flooding concerns coexisting with dependence on gravity in the case of absorption columns. GLC combines polymeric membrane separation and chemical absorption, using a physically and chemically active solvent for selective CO_2 removal. However, the new aspect is that there is a membrane standing as a physical barrier against the unnecessary mixing of gas and liquid phases. Other advantages of GLC are: (i) it allows independent manipulation of liquid and gas flows; (ii) offers larger area of gas–liquid interface; (iii) modularity ; (iv) it exhibits flexibility to increase/decrease operational scales; (v) no dew-point conditioning of the gas feed is necessary; and (v) angular indifference allowing horizontal or vertical operational arrangements. Figure 14 sketches a typical process flowsheet for operation of a GLC unit capturing CO_2 from a NG feed.

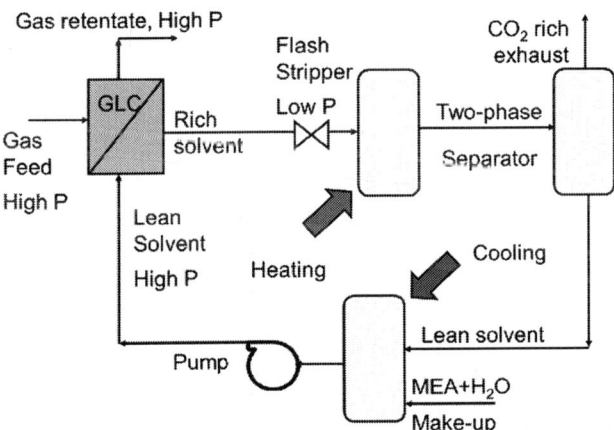

Figure 14: Typical process flowsheet for CO_2 capture with gas-liquid contactor (GLC).

Here the GLC operates t high NG pressure, but the rich solvent has to be regenerated in a low pressure stripper where CO_2 is released by the action of heating and low pressure as occurs in the second regeneration column shown in Figure 6.

A model for stationary gas-liquid hollow-fiber contactor was proposed by de Medeiros et al. (2013a) for separating CO_2 from CO_2 rich natural gas. This model is based on the AGWA theory discussed in the sub-section 3.1 (de Medeiros et al., 2013b). The model assumes a hollow-fiber contactor with co-current compressible flows of both permeate and retentate, where the permeate corresponds to the inner space inside the hollow-fibers. Permeate and retentate are separated by the membrane. The permeate is supposed in two-phase flow because the trans-membrane flux of methane will support the maintenance of a gas phase in the inner permeate flow. The permeate is, in fact, a two-phase reactive flow in continuous reactive vapor-liquid equilibrium (RVLE) because CO_2 is in reactive vapor-liquid equilibrium with water and amine via a set of chemical equations similar to Eq. (9) (without the H_2S chemical equations if the GLC is designed for CO_2 capture only).

The model of de Medeiros et al. (2013a) is based on 1-dimensional axial geometry of the GLC. Species mass balances and momentum/energy balances are written for both RVLE permeate and gas retentate. All properties of gas and liquid phases are calculated with PR-EOS. The trans-membrane flux terms are written as products of permeancy coefficients and a driving force term. Each species has a permeancy coefficient for a given membrane and given temperature conditions. The driving force term is expressed as a difference of component fugacities between the retentate and RVLE permeate sides. The major difficulty encountered in the GLC model is to represent the RVLE two-phase flow in the inner membrane space. This is accomplished by solving the AGWA VLE model (de Medeiros et al., 2012b) described in Table 1, along the entire path of the permeate in the inner space of the hollow-fibers. As an example, consider the equilibrium map in Figure 15.

This map was calculated assuming that 1 mol of natural gas (with 16.7%mol CO_2 +82.3%mol CH_4) is contacted with 1.2 mol of liquid solvent containing 14.5%mol MEA +14.5%mol MDEA +71%mol H_2O. Only the chemical reactions of CO_2 absorption by MEA (Eq. 9a) and MDEA (Eq. 9b) are considered. Each point in Figure 15 represents the resulting equilibrium vapor phase mol fraction of CO_2 (*YCO2*) versus $T(°C)$ X P (bar) under reactive VLE. Clearly, the locations with low *YCO2* (blue) are dominated by absorption phenomena, while those with high *YCO2* (red) are dominated by stripping phenomena. This kind of inner

RVLE solver is a key element in the construction of the GLC model of de Medeiros et al. (2013a).

The CO_2 Transportation Cycle

An efficient and reliable transportation system is required to displace enormous quantities of captured CO_2 to their destination site at appropriate geological formations that are capable to accommodate hundreds of billions or trillions of Nm^3 of CO_2 under stable and secure conditions. Clearly a not especially large size thermoelectric coal plant can produce something like 1 million of metric tonnes (1Mt) of CO_2 per year or about 500.10^6 Nm^3/y. Such huge capacities can only be attended by large pipeline systems operating with dense compressed fluid (liquid or dense supercritical CO_2), because the other existing alternatives – road, railroad and barge transport – simply cannot cope with dense fluid pipelines in terms of unitary cost and capacity of transportation.

In other words, despite of the existence of many options for transporting compressed (gas or liquid) CO_2 – including highway tankers, railway tankers, ships, and pipelines – it is evident that the impressive volumes that must be transported dictates that only pipelines working at high pressures, high densities and high capacities are suitable for this service. For instance, 2-3Mt/y of CO_2 have to be transported to dispose of the entire production of a single 500MW coal fired power plant. This corresponds to transporting 230-350t/h of CO_2, just to service a single, medium-sized, client. Thus, only a network of large-scale pipelines could provide viable overland transport of such massive flow rates of CO_2. Presently, about 50Mt/y of CO_2 (equivalent to the output of 16 coal-fired power plants) are transported by 3100km of CO_2 pipelines, mainly for EOR processes in the USA and Canada (de Medeiros et al., 2008). The best example is the 808 km long, 30" diameter, Cortez Pipeline that transports 13Mt/y of CO_2 from highlands in Colorado to oilfields in Texas, USA.

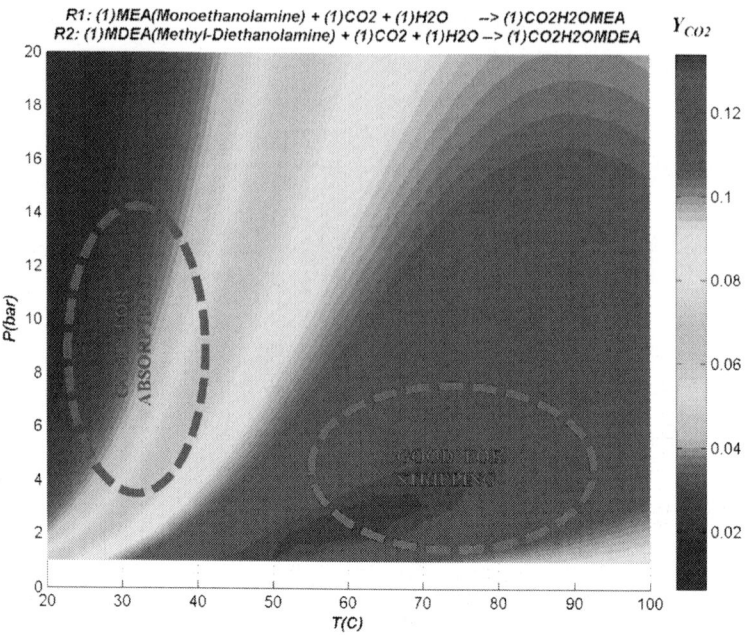

Figure 15: Equilibrium vapor phase mol fraction of CO_2 ($YCO2$) versus $T(°C)$ X P (bar) under reactive VLE: 1mol of a 16.7%mol CO_2 natural gas is contacted with 1.2 mol of a liquid solvent with 14.5%mol MEA, 14.5%mol MDEA and 71%mol water.

Based on historical capital and O&M data for a 480-km long CO_2 pipeline without booster compressors, McCoy (2008) projected a fixed O&M coefficient of $3,250/y/km for CO_2 pipelines. Considering a horizontal pipeline without appreciable elevation changes and an annualized fixed cost of 15% of capital, McCoy estimated the total unitary cost of CO_2 transportation as only $1.16 per tonne of CO_2 per 100km. Based on a Monte Carlo sensitivity analysis, McCoy (2008) determined a range of $0.75 to $3.56 per tonne of CO_2 per 100km for this cost, recommending the median value of $1.65 per tonne per 100km as a suitable estimate for investment decisions.

The design of CO_2 pipelines depends on reliable compressible flow models for dense compressible fluid near critical conditions. This model should account for thermal compressibility effects inside the fluid, i.e., temperature increases (decreases) during downhill (uphill) flow due to gravity compression (expansion).

In the same way, external heat transfer and elevation effects must be allowed. The extremely high compressibility of CO_2 near its critical state at 31°C (shown in Figures 4 and 5), leads to potential abrupt changes of velocity due to abrupt changes of density as the fluid compresses (decompresses) near the critical state. In other words, any candidate model for high capacity CO_2 pipelines must be able to calculate thermodynamic properties of dense supercritical CO_2 near its critical transition with accuracy. Such a CO_2 pipeline model has been proposed by de Medeiros et al. (2008) by solving rigorous species mass balances and energy/momentum balances along the pipeline with all thermodynamic properties given by PR – EOS or SRK – EOS.

This model is demonstrated in the simulation of a sub-sea CO_2 pipeline (Figures 16, 17 and 18) for transportation of 20 MMSm³/d of CO_2 from onshore plant to five EOR wellheads 2100m deep, 380km from the coast. The 20" pipeline extends 380km from west to east and 320km from south to north with 700 km of length.

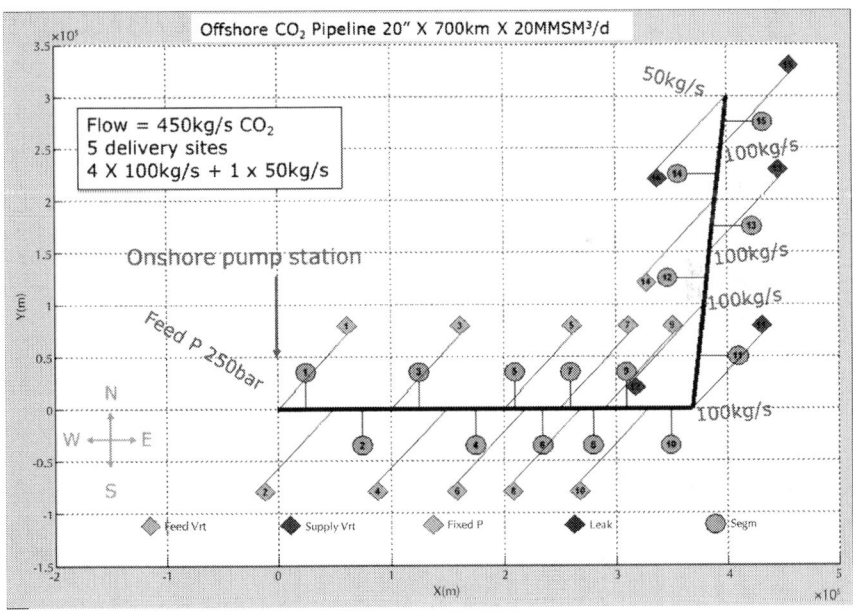

Figure 16: Hypothetical 20"X 700 km sub-sea pipeline for 20MMSm³/d of liquid CO_2 (≈450kg/s) at 250 bar from onshore plant to five offshore EOR wellheads (4X100kg/s + 1X50kg/s) 2100m deep, 380km from the coast.

As seen in Figure 17, a problematic factor was inserted between the continental shelf and the continental slope, namely, a big rift 1500m deep lies in the pipeline route. Rifts are not common in Santos Basin, Brazil, but they exist in the Norway arctic coast (Pettersen, 2011).

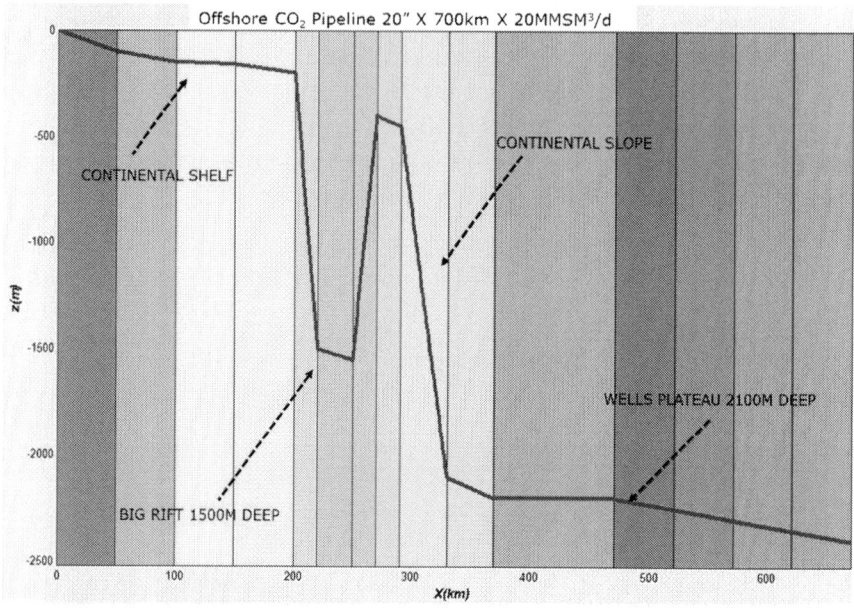

Figure 17: Elevation profile for the hypothetical sub-sea CO_2 pipeline from onshore plant to five EOR offshore wellheads 2100m deep, 380km from the coast.

THE CO_2 UTILIZATION CYCLE

Although CGS has been regarded worldwide as a mitigation technology, it deals with CO_2 as a waste with an energy and an economic penalty for its disposal (Armstrong, 2012). Rather than treating CO_2 as a waste, carbon dioxide utilization (CDU) recognizes it as a raw material in chemical process to produce high added-value carbon containing products. It is also worth noting that the CDU is a complementary technology to CGS, not a competing technology, adding value to a process and thus it may help balance the costs of CGS.

CO_2 Utilization: A Process Systems Engineering Vision

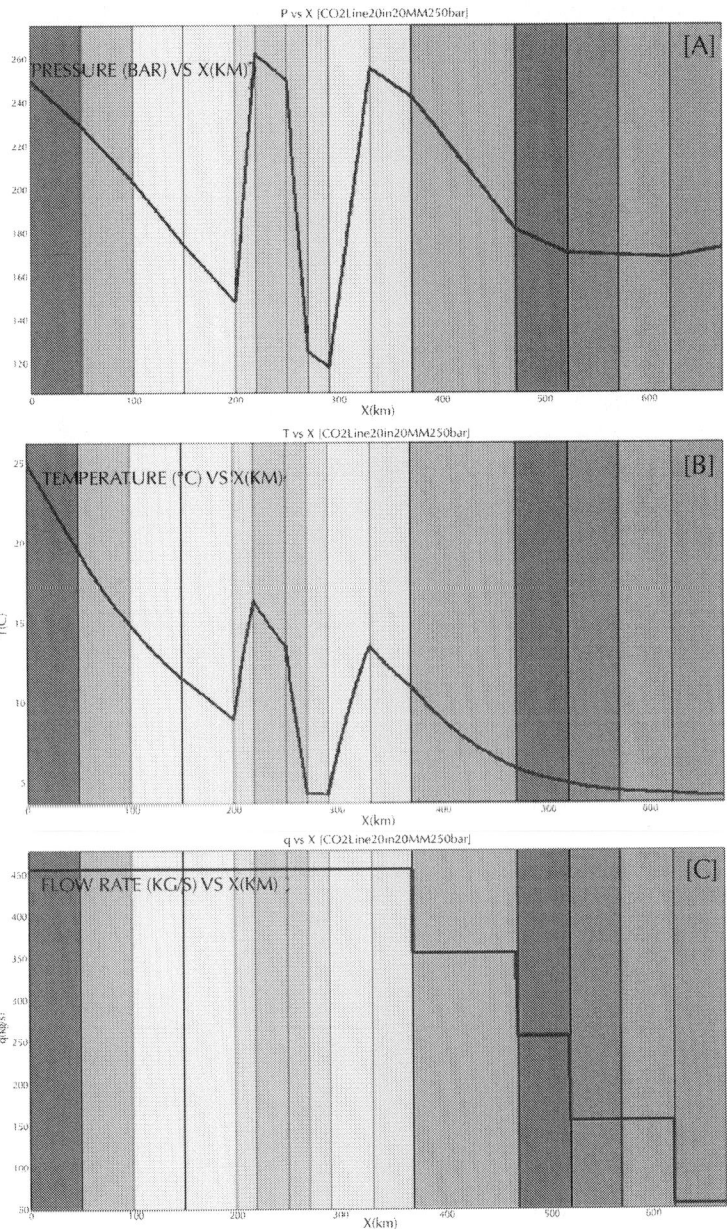

Figure 18: Calculated profiles for the hypothetical 20"X 700 km sub-sea pipeline with 20MMSm^3/d ($\approx$ 450kg/s): [A] pressure (bar); [B] temperature (°C); [C] mass flow rate (kg/s).

While CO_2 has broad industrial application as solvent (supercritical CO_2, fire extinguishers) and in the food industry, it has disadvantages as a chemical raw material due to its low reactivity and few reactions are thermodynamically feasible. Furthermore, each potential use of CO_2 as reactant has an energy requirement that needs to be determined and must not exceed the CO_2 utilized and, although the utilization of CO_2 has been subject of research since before 1970´s, there is much research still needed for CO_2 activation. Moreover, the utilization of CO_2 to cause an effective reduction in its emission into the atmosphere, must observe certain guidelines (Aresta, 2010): (i) the new process must reduce the overall CO_2 emissions; (ii) it mustbe less energy - and material - intensive with respect to the on–stream processes that it aims to replace; (iii) the new process must reduce the overall CO_2 emissions; (iv) it must employ safer and more eco-friendly working conditions; (v) it needs to be able to operate on a commercial scale and (vi) it must be economically viable.

According to Song (2006), the global market for CO_2 is estimated to be \$3.2 billion/year in 2003. Carbon markets across the world were valued at 96 billion euros (\$122.28 billion) in 2011 (Reuters Agency, 2012). Utilization of CO_2 by the chemical industry exists for more than one century as, for instance, the synthesis of urea (50 Mt/y) (Aresta, 1999) (Eq. 11), salicylic acid (Song, 2006) (Eq. 12)and inorganic carbonates (20Mt/y) (Aresta, 1999).

$$CO_2 + 2\,NH_3 \longrightarrow H_2N-CO-NH_2 + H_2O$$

$$C_6H_5OH + CO_2 \longrightarrow C_6H_4(OH)(COOH)$$

According to Song (2006), the worldwide production of urea in 2002 was about 110 million metric tonnes, which contains 51.8 million metric tonnes of nitrogen with an estimated value of US\$11.5 billion. This corresponds to about 81 million metric tons of CO_2, and 22 million metric tonnes of carbon.

Thermodynamic and Chemical Considerations of CO_2 Conversion

Chemical reactions are driven by the difference in Gibbs free energy between the products and reactants at certain conditions. The obstacle for utilizing CO_2 as feedstock to industrial processes is its low energy level - CO_2 is a highly stable molecule. Consequently, a substantial input of energy, effective reaction conditions, and often catalysts, are necessary for its chemical conversion. In other words, many reactions for CO_2 conversion involve positive change in enthalpy, requiring an energy input. There are many large-scale chemical industrial processes that are operated based on endothermic reactions in the chemical industry (e.g., ammonia Haber process).

Song (2006) states properly that it is more energy-demanding if one were to use only CO_2 as a single reactant, but it becomes easier in thermodynamically terms if CO_2 is used as a co-reactant with another substance that has higher Gibbs free energy, such as CH_4, graphite and H_2. Song (2006) illustrates this trend by the change in the reaction heat for reactions with CO_2 as the single reactant (Eq. 13) and with CO_2 as a co-reactant (Eq. 14).

$$CO_2 \longrightarrow CO + \tfrac{1}{2} O_2 \qquad \Delta H^0 = +293 \, \frac{kJ}{mol\,CO_2}$$

$$CO_2 + H_2 \longrightarrow CO(g) + H_2O \qquad \Delta H^0 = +51 \, \frac{kJ}{mol\,CO_2}$$

Therefore, energy input is necessary to transform CO_2 into chemicals. Four methods are possible: (i) reaction with high-energy molecules (e.g., ethylene oxide, H_2, unsaturated compounds and organometallic compounds); (ii) reaction with low energy oxidized compounds (e.g., organic carbonates), (iii) shifting chemical equilibrium towards products (via removal of a reaction product) and (iv) supplying physical energy (e.g., light or electricity) (Sakakura et al., 2007). The appropriate selection of chemical reactions makes it possible to obtain a negative Gibbs energy change.

As the carbon of the carbonyl group has an electron deficiency, CO_2 has great affinity for nucleophilic compounds and electron donors, i.e., as an anhydrous carboxylic acid, it promptly reacts with basic compounds. For instance, organometallic compounds, such as Grignard compounds, react promptly with CO_2 even at low temperatures. Reactions with CO_2 can be divided into two groups: (1) formation of a carboxylic group via a nucleophilic attack and (2) oxidative cycle addition yielding a ring of 5 members (Sakakura et al., 2007).

A relevant aspect to be considered is that utilization of CO_2 as feedstock does not necessarily contribute to the mitigation of greenhouse effects, even though CO_2 stands as a green reactant in many cases (Sakakura et al., 2007). Three points are hence relevant:

The chemical (or biochemical) fixation of CO_2 does not necessarily imply in reducing CO_2 emissions as its transformation requires energy, both to drive reaction (high temperatures and pressures) and separate products (separation occurs mainly at low pressures and, hence, recycling unreacted CO_2 to the reactor will require recompression at the expense of high energy input);

The energy demand of the world is order of magnitude higher than the amount of CO_2 fixed by chemical utilization of CO_2, and

In the critical phase of its life cycle, organic chemicals will emit CO_2. Nevertheless, the relevance of CO_2 as raw material stands for being a renewable feedstock, substituting the conventional fossil based routes

Among other utilization, CO_2 is currently used as supercritical solvent, refrigerant fluid, beverage carbonation agent, inert medium (such as fire extinguisher), pressurizing agent, neutralizing agent, gas for greenhouses, "inerting" applications to inhibit unintended chemical reactions, welding (preventing atmospheric oxygen from reacting with molten metal), food processing (suppressing aerobic bacterial activity for preservation in processes like pneumatic conveying or food storage). In any of these applications for inerting, carbon dioxide serves as a cover against atmospheric oxygen and is thus implicitly released into the atmosphere. The carbonation of beverages accounts for around 1.0×10^6 t CO_2/y. Nevertheless, they do not constitute CO_2 sink as it is ultimately released to the atmosphere or remains in a closed loop (Ormerod et al., 1995). Hence, this study does not review such utilizations.

Supply Chain Considerations of CO_2 Conversion

Bayer (2013) estimates that the chemical industry has over 40,000 final chemicals, produced from approximately 400 intermediate chemicals, derived from ~40 basic chemicals that, in turn, are based on 4 classic feedstocks: petroleum, natural gas, coal and biomass. The Company expands the set of feedstock with the inclusion of CO_2.

For the near and middle term time-period, i.e., next one or two decades, it is reasonable to assume that presently dominant technologies (from an economic standing point) will persist and, consequently, expanding economies´ demand of energy will be met by present technologies. Consequently, GHG emission of chemical processes will expand. This same approach leads to a transition scenario to a low-carbon economy equally dominated by presently installed infrastructure.

Hence, the CO_2 utilization cycle is likely to rely on commercially mature technologies or on technologies presently in large-scale pilot or demonstration plants as *Bridge Technologies*. Therefore, only peripheral technological advances are expected in these technologies such as process intensification, enhanced selectivity and activity of catalysts, and process optimization with increased mass and energy integration.

It is worth noting that 5 chemical commodities, ammonia, methanol, ethylene, propylene and BTX dominate energy consumption and GHG emissions in the chemical industry (IEA, 2013). In the conception of co-processing of fossil feedstock, with biomass and CO_2, conversion routes to produce ammonia, methanol and olefins (e.g., ethylene and propylene) are considered (Figure 2). Although not included in Figure 2, catalytic fast pyrolysis of biomass can lead to the key aromatic compounds, Benzene, Toluene, and Xylene (BTX), with generation of paraxylene from the BTX and subsequent conversion to Purified Terephthalic acid (PTA) and PolyEthylene Terephthalate (PET).

Synthesis gas (SYNGAS), a mixture of hydrogen, carbon monoxide and CO_2, is a versatile intermediate feedstock used in the production of a number of hydrocarbons such as methanol, ammonia, synthetic hydrocarbon liquids, and as a source of pure hydrogen and carbon monoxide. Applications of these products range from petrochemical feedstock to fuels.

According to Rostrup-Nielsen and Christiansen (2011), trends in the use of SYNGAS are dominated by the conversion of inexpensive remote natural gas into liquid fuels ("gas to liquids" or "GTL") and by a possible role in a future "hydrogen economy" mainly associated with the use of fuel cells. Some relevant synthesis to the chemical industry are:

SYNGAS to Ammonia: Ammonia serves as a building block in many pharmaceuticals, fertilizers, ethanolamines, urea and cleaning products, as well as an anti-microbial agent in food processing. 50% of the world´s food production relies on ammonia fertilizers.

SYNGAS to Methanol: The main use for methanol is to produce other chemicals; about 40% is converted to formaldehyde, and further processed into plastics, plywood, paints, explosives and textiles. It is also used in anti-freeze, solvents, and fuels, and can serve as energy carrier.

SYNGAS to Hydrogen: Hydrogen generation is one of the largest energy-consuming steps in the production of the crucial chemical precursors of ammonia and methanol.

SYNGAS to Synthetic Fuels: Liquid hydrocarbons exhibit an excellent volumetric energy density and offer various opportunities for storing electric energy (Kaiser et al., 2013). Kaiser et al. (2013) point generation of SYNGAS by reverse water-gas shift (RWGS) at elevated temperatures as the first step, followed by Fischer-Tropsch (FT) synthesis. If CO is substituted by CO_2, less synthetic fuels are formed, the water-gas shift is repressed, and methane selectivity increases.

Co_2 to Syngas

SYNGAS is a toxic, colorless and odorless mixture. Its efficient commercial production is gaining significant attention worldwide (Raju et al., 2009) as it is a versatile feedstock to produce a variety of fuels and chemicals.

Almost any carbon source ranging from natural gas and oil products to coal and biomass can be used in the SYNGAS production. The lowest cost routes for its production, however, is natural gas (Spath and Dayton, 2003), which is also the cleanest of all fossil fuels. Furthermore, steam methane reforming is a well-established process for the production of SYNGAS and hydrogen (Gangadharan et al.,

2012). The H_2/CO ratio varies over a wide range, depending on the primary feedstock and technology employed. Particular SYNGAS ratios are required depending on the chemical product desired, therefore creating flexibility for the chemical industry.

In the twofold context of avoiding emissions and standing as a renewable feedstock, carbon dioxide has been investigated as raw material in SYNGAS production. The new technologies involves CO_2: (i) reforming processes using a hydrocarbon (methane, typically) as reducing agent; (ii) using CO_2 as a co-reactant with hydrogen in the catalytic reverse water gas shift (RWGS); (iii) thermocatalytic (solar assisted) routes; (iv) electro- or photo-catalysis; (v) plasma processes, and (vi) bio-processes, e.g., by hybrid enzyme-nanoparticle systems, bioelectrochemical reduction or using a biomass char and a catalyst such as Ni/Al_2O_3.

CO_2 Reforming of CH_4 (Dry Reforming): The dry (carbon dioxide) reforming of methane has been of interest for a long time, dating back to as early as the 1920s, and was first proposed by Fischer and Tropsch (1928), but it is only in recent years that interest in it has rapidly increased for both environmental and commercial reasons (Zhang et al., 2003). Its name derives from the fact that CO_2 replaces steam in the conventional steam methane reforming process (Hartley & Tam, 2012). This reaction can be represented as

$$CH_4 + CO_2 \longrightarrow 2CO + 2H_2 \qquad \Delta H = +247 \frac{kJ}{mol}$$

According to Hartley and Tam (2012), dry reforming utilizing CO_2 produces synthesis gas with higher purity and lower H_2/CO ratio than either partial oxidation or steam reforming. The produced SYNGAS has an H_2/CO ratio of unity without further post-reformer reactions (Zhang et al., 2003). The interest in this reforming route in recent years (Treacy and Ross, 2004, Shi et al., 2013) is due to two main reasons: (i) it produces SYNGAS with a H_2/CO molar ratio that is suitable for a variety of products including Fischer–Tropsch fuels and (ii) the reaction consumes two types of greenhouse gases, CO_2 and CH_4 (Zhang et al., 2003, Gangadharan et al., 2012). Moreover, SYNGAS production stands as the most promising alternative of CO_2 utilization as it presents flexibility of using installed infrastructure to the manufacture of important chemical commodities.

The biggest limitation to the dry reforming process appears to be the availability of a suitable catalyst. The high temperatures required to reach high conversions, due to the endothermic nature of the process, contribute to carbon deposition (both CO_2 and CH_4 give off carbon deposits), and a catalyst capable of operating at such severely deactivating conditions has not been well developed (Zhang et al., 2003, Shi et al., 2013). According to Shi et al (2013), from the viewpoint of GTL industry, developing a catalyst for CO_2 reforming of CH_4 is a challenge, because the catalyst must exhibit very high production rates to render the GTL methane reformer as small as possible. Nevertheless, progress in the development of suitable catalysts with higher activities and optimized lifetime stabilities have been reported (Bradford and Vannice, 1999, Souza and Schmal, 2003; Zhang et al., 2003, Ginsburg et al., 2005; Kahle et al., 2013; Shi et al., 2013; Zheng et al., 2013; Edwards, 1995; Wurzel et al., 2000; Nagaoka et al., 2001; Li et al., 2004). Nevertheless, there is still no process for the CO_2 reforming currently considered to be commercially feasible. However, a variation of dry reforming has been used industrially (Hartley and Tam, 2012). The CALCOR process (Teuner, 1985, Kurz and Teuner, 1990, Teuner et al., 2001) involves dry reforming of methane, optimized to reduce the hydrogen content of the product gas. Furthermore, hydrogen separation by membrane permeators produces hydrogen gas that combusts a fuel (e.g., methane) producing pure carbon monoxide. The SPARG process (promotion by poisoning) is also a dry reforming reaction process (Gunardson, 1998; O'Connor and Ross, 1998; Rostrup-Nielsen, 2006). The active catalytic sites are blocked by poisoning the feed gas with hydrogen sulfide (H_2S). The adsorption of sulfur at the catalytic sites is favored over carbon growth. The SPARG process is claimed to produce high CO content SYNGAS.

Also combined CO_2 and steam reforming systems have been operational in the industry for a number of years (Gangadharan et al., 2012). By choosing the right proportions between CH_4, water and CO_2 (3/2/1), the combination of steam and dry reforming of methane can generate SYNGAS with a H_2/CO ratio of 2, ideal, for example, for the synthesis of methanol (Rostrup-Nielsen and Christiansen, 2011;Olah et al., 2009). This combination of steam and dry reforming was named bi-reforming.

Bi-reforming could be advantageous in the use of various natural gas sources even these containing substantial amounts of CO_2. Some natural gas as well as biogas sources contain CO_2 concentration up to 50–70%. Bi-reforming can also be used to recycle CO_2 emissions from sources such as flue gases from fossil fuel (coal, petroleum, natural gas, etc.), burning power plants, exhaust of cement factories, among other industries (Olah et al., 2013). This reaction can be represented as

$$\text{Steam reforming: } CH_4 + H_2O \longrightarrow CO + 3H_2 \quad \Delta H = +206.3 \frac{kJ}{molCO_2}$$

$$\text{Dry reforming: } CH_4 + CO_2 \longrightarrow 2CO + 2H_2 \quad \Delta H = +247.3 \frac{kJ}{molCO_2}$$

$$\text{Bi-reforming: } 3CH_4 + 2H_2O + CO_2 \longrightarrow 4CO + 8H_2$$

Bi-reforming is adaptable for reforming varied natural gas (containing hydrocarbon homologues) and CO_2 sources, e.g., shale gas (Olah et al., 2013):

$$3C_nH_{(2n+2)} + (3n-1)H_2O + CO_2 \longrightarrow (3n+1)CO + (6n+2)H_2$$

Numerous authors (Ashcroft et al., 1991; O´Connor and Ross, 1998; Wang et al, 200;,Jarungthammachote, 2011) have studied a similar idea, which combines dry reforming with partial oxidation. The idea again being that the combination helps overcome the endothermic requirement of dry reforming with the exothermic nature of partial oxidation, resulting in lower total energy consumption. In addition, it allows altering the H_2/CO ratio by controlling the extent to which each reaction takes place (Hartley and Tam, 2012). The combination of exothermic and endothermic reactions is called autothermal reaction (ATR). The ATR technology requires addition of CO_2 or CO_2-rich gas, in order to adjust the SYNGAS composition to the desired H_2/CO ratio.

$$\text{Dry reforming: } CH_4 + CO_2 \longrightarrow 2CO + 2H_2 \quad \Delta H = +247.3 \frac{kJ}{mol}$$

$$\text{Partial Oxidation of Methane: } CH_4 + 0.5O_2 \longrightarrow CO + 2H_2 \quad \Delta H = -35.6 \frac{kJ}{mol}$$

The combined Dry Reforming and Partial Oxidation is hence:

$$CH_4 + 0.5CO_2 + 0.25O_2 \longrightarrow 1.5CO + 2H_2 \quad \Delta H = +211.7 \frac{kJ}{mol} \qquad (22)$$

Integrating steam reforming and partial oxidation with CO_2 reforming could reduce or eliminate carbon formation on reforming catalyst, thus increasing catalyst life and process efficiency. Therefore, the tri-reforming, a synergetic combination of CO_2 reforming, steam reforming, and partial oxidation of methane in a single reactor for effective production of industrially useful SYNGAS (Song, 2006) could solve two important problems encountered in individual processing. Incorporating oxygen in the reaction generates heat *in situ* that could increase energy efficiency; oxygen also reduces or eliminates carbon formation on the reforming catalyst. The tri-reforming can be achieved with natural gas and flue gases using the waste heat in power plants and the heat generated *in situ* from oxidation with the oxygen that is present in flue gas (Zhou et al., 2008; Zangouei et al., 2010; Moon et al., 2004).

The tri-reforming process is presented in Eqs. (23) to (26) (Song and Pan, 2004):

Steam reforming: $CH_4 + H_2O \longrightarrow CO + 3H_2 \quad \Delta H = +206.3 \frac{kJ}{mol}$

Dry reforming: $CH_4 + CO_2 \longrightarrow 2CO + 2H_2 \quad \Delta H = +247.3 \frac{kJ}{mol}$

Partial Oxidation of Methane:

$$CH_4 + 0.5O_2 \longrightarrow CO_2 + 2H_2 \quad \Delta H = -35.6 \frac{kJ}{mol}$$

$$CH_4 + 0.5O_2 \longrightarrow CO + 2H_2O \quad \Delta H = -880 \frac{kJ}{mol}$$

Song (2006) reports experimental and computational results to support that tri-reforming produces SYNGAS with desired H_2/CO ratios (1.5–2.0) and eliminates carbon formation in the CO_2 reforming of CH_4. Song (2006) suggests that tri-reforming is especially suited to using CO_2 in concentrated sources without prior CO_2 separation, as in non-conventional (low-quality CO_2-rich) natural gas, and has been demonstrated in pilot scale in Korea.

In general, produced SYNGAS from methane reforming is converted catalytically *in situ* via one of two main routes. The first is to use Fischer-Tropsch synthesis, a process that catalytically converts SYNGAS to hydrocarbons of varying molecular weights. The second is methanol synthesis. The latter has better atomic economy, since the oxygen atom in CO is included in the product and CO_2 can be blended into SYNGAS as a reactant. However, production of methanol is very inefficient in this reaction: only 10-15% one pass conversion typically at 5.0-10.0 MPa and 523-573 K, due to the severe thermodynamic limitations of this exothermal reaction ($CO+2H_2 \rightarrow CH_3OH$) (Shi et al., 2013).

Finally, CO_2 reforming of methane can also be used as a chemical energy storage alternative and an energy transmission system (Richardson and Paripatyadar, 1990, Levitan et al., 1991; Levy et al., 1993). According to Zhang et al. (2003), in this system, solar energy is used to drive the endothermic forward reaction, and the energy thus stored can be transported via pipelines such as SYNGAS and liberated at will by the reverse reaction at any location or time. The highly endothermic reaction could be an option to store solar energy in hot regions (Zhang et al., 2013).

Reverse Water Gas-Shift (RWGS): The reverse water gas shift (RWGS) reaction has been known from over two centuries and is a well-researched and understood process for SYNGAS ratio alteration (Hartley and Tam, 2012). In fact, both the water gas shift (WGS) and the RWGS reactions are mostly used in combination with reforming of hydrocarbons to adjust the H_2/CO ratio, as shown in Eq. (27)(Song, 2006). Depending on the reaction conditions, the equilibrium for the WGS can be pushed in either the forward or the reverse direction. Efforts to explain the RWGS reaction mechanism are reported (Goguet et al., 2006, Meunier et al., 2007, Wang et al., 2013), and two main mechanisms have been proposed: the *redox* mechanism and the *associative formate* mechanism. The reversibility of the WGSR is important in the production of ammonia, methanol, and Fischer-Tropsch synthesis where the ratio of H_2/CO is critical. Many industrial companies exploit the RWGS reaction as a source of the synthetically valuable CO from cheap CO_2. In fact, catalytic RWGS reaction is the main route to produce SYNGAS from CO_2.

$$H_2 + CO_2 \longleftrightarrow CO + H_2O \quad \Delta H = +51 \frac{kJ}{mol}$$

RWGS provides a source of hydrogen at the expense of carbon monoxide, which is important for the production of high purity hydrogen. This is a mildly endothermic reaction, as shown in Eq. (27).

Although high temperature reactions are effective for obtaining a high conversion, WGS reaction is an equilibrium-limited reaction that exhibits decreasing conversion with increasing temperature. In order to take advantage of both the thermodynamics and kinetics of the reaction, industrial scale WGS reaction is conducted in multiple adiabatic stages consisting of a high temperature shift (HTS) followed by a low temperature shift (LTS) with intersystem cooling (Byron, 2010). The initial HTS takes advantage of the high reaction rates, but is thermodynamically limited, which results in incomplete conversion of carbon monoxide and a 2-4% carbon monoxide exit composition. To shift the equilibrium towards hydrogen production, a subsequent low temperature shift reactor is employed to produce a carbon monoxide exit composition of less than 1% (Byron, 2010). A catalyst is required under these conditions because of the lower reaction rate at low temperatures. The RWGS reaction uses a variety of catalysts, including palladium, platinum on titania, copper, cobalt with manganese/zinc oxide and rhodium with ceria (Tanaka et al., 2003, Saito and Murata, 2004, Meunier et al., 2007). Many research groups are looking at copper as a catalyst due to its effectiveness and its relatively low cost (Armstrong et al., 2013). However, there has been renewed interest in the WGSR at extreme temperatures, because of recent advances in high-temperature materials for hydrogen separation membranes (Bustamante et al., 2002).

Co_2 to Methanol

Methanol is one of the most important commodity chemicals as it is used as a raw material in several intermediate chemicals and end uses. Methanol is produced industrially from SYNGAS from natural gas or coal mainly containing CO, H_2 and a small amount of CO_2 in presence of a catalyst. Nevertheless, direct CO_2 hydrogenation has also been reported. Other non-conventional routes are electro- or photoprocesses, as well as the use of enzymes. The importance of methanol synthesis is demonstrated by widespread scientific publications of various reaction routes (Razali et al. 2012), and the development of several pilot plants to use waste carbon dioxide for methanol production.

Among the new technologies, in terms of potential for application, the CO_2 catalytic hydrogenation to methanol appears to have the highest degree of commercialization. It may be already commercially interesting when cheap sources of renewable H_2 are available, or to store excess electrical energy, as an alternative to actual systems. It is estimated that this reaction could reach the industrial stage in less than five years. This development would be pushed by experience in pilot or pre-commercial industrial plants, such as the Mitsui Chemicals Inc.'s plant (pilot in Japan capable of producing 100 t of methanol per year, and large unit expected in Singapore) and a plant by Carbon Recycling International (installed at the end of 2010) (Quadrelli et al., 2011). Mitsui's pilot plant uses CO_2 from an ethylene production plant of Osaka Works Petrochemical Complex (ADEME, 2010). It synthesizes methanol by CO_2 hydrogenation and the simultaneous water gas shift reactions. The process claims 96% selectivity (Hartley and Tam, 2009). Carbon Recycling International is capable of producing 3000 t/y of methanol (ADEME, 2010). This unit has a capacity of about 10 t of methanol from 18 t of CO_2 (Carbon Recycling International, 2009; Van-Dal and Bouallou, 2013), with CO_2 from the Svartsengi geothermal plant and an aluminum production plant. Hydrogen is generated from the electrolysis of water using a renewable source of electricity.

Methanol from SYNGAS: Synthesis gas composed of the proper ratio of hydrogen, carbon monoxide and carbon dioxide is converted to methanol. Alternatives paths to methanol are via CO from RWGS reacting with hydrogen according to Eq. (28) and via CO_2 being hydrogenated following Eq. (29).

$$CO + 2H_2 \longrightarrow CH_3OH \quad \Delta H = -90.6 \frac{kJ}{mol}$$

$$CO_2 + 3H_2 \longrightarrow CH_3OH + H_2O \quad \Delta H = -49.5 \frac{kJ}{mol}$$

From Eq. (28), production of methanol involves SYNGAS production as intermediate stage. Hence, two steps are required for the manufacture of methanol: reduction to SYNGAS and reaction to form methanol. There are process variations for implementing the sequence. Before being sent to the methanol production unit, the SYNGAS must thus be subjected to the WGS reaction to enhance its hydrogen content. Alternatively, H_2 from other sources can be added. Recent efforts have been aimed at production of methanol in a one-step process without intermediate formation of SYNGAS. Homogeneous or heterogeneous catalysts are typically preferable. The conventional process occurs at relatively low pressures (5 to 10 MPa) and 210 to 350 °C employing a $Cu/ZnO/Al_2O_3$ catalyst.

Catalytic Hydrogenation Conversion of Carbon Dioxide to Methanol: The most direct and studied route to methanol from CO_2 is the catalytic conversion of CO_2 with hydrogen. Carbon dioxide hydrogenation to methanol is a relatively mature process. The main issue is the cost (and associated carbon footprint) of the H_2 necessary for the reaction. Any available energy source (alternative energies such as solar, wind, geothermal, and atomic energy) can be used for the production of needed hydrogen and chemical conversion of CO_2. The process can use lower operational pressures of 3 MPa at 240 °C. This direct CO_2 hydrogenation exhibits low conversions resulting in high volumes of recycled gas. Literature indicates that methanol is synthesized following a 3:1 hydrogen to carbon dioxide stoichiometry using catalysts of copper oxide, zinc oxide, incorporating either titania, aluminum oxide, chromium oxide and alternatively lanthanum or gallinium (Lachowska and Skrzypek, 2004, Lee et al., 2004, Stoczynski et al., 2004).

An alternative approach to the use of solid catalysts and a gas phase process is to employ the so called low-temperature methanol synthesis (LTMS) (Dixneuf, 2011). LTMS is based on the catalytic hydrogenation of methanol to formic acid (HCOOH) with subsequent etherification to methanol formate (alternative to methyl formate from SYNGAS), followed by hydrogenation of formate to two methanol molecules using Pincer-type ruthenium(II) catalyst (Balaraman et al, 2011; Dixneuf, 2011; Huff and Sanford, 2011). A liquid-phase allows CO_2 and H_2 conversion to methanol of about 95% with very high selectivity in a single pass (Olah, 2009). Waugh (2012) has published a review on catalytic methanol synthesis which includes the use of carbon dioxide as a feedstock.

Photoreduction of CO_2 to methanol: Photoelectrochemical reduction of carbon dioxide or photocatalysis generally uses semiconductors to promote reaction in the presence of sun light. The semiconductor is used as a catalyst to absorb solar energy and generate electrons and protons needed for the reduction of carbon dioxide. While hydrogenation of carbon dioxide requires high temperature and high pressure conditions, photocatalysis carries out under relatively mild conditions with advantageous energy input – sun light – a continuous and readily available source (Le, 2009).

Considerable research effort has been made on CO_2 activation by visible light photocatalysts due to the natural abundance of sunlight. Nevertheless, the efficient photoreduction of CO_2 with H_2O remains one of the most challenging tasks of environmental catalysis.

CO_2 can be reduced in water vapor or solvent by photocatalysts such as TiO_2 and ZnS. Eq. (30) describes the overall reaction.

$$CO_2 + 2H_2O \xrightarrow{h\nu} CH_3OH + 3/2\, O_2$$

Due to the high energy requirements, this method is often combined with electrochemical methods via photoelectrocatalysis to drive the reaction (Hu et al., 2013). The catalysts traditionally used are transition metal complexes, TiO_2, ZnO, CdS, and functionalized metal surfaces (Yamashita et al., 1998;Kuwabata et al., 1994). A wide variety of CO_2 photoreduction has been achieved on the surface of TiO_2 under UV irradiation. The yield of photoproducts can be changed substantially under different experimental conditions such as UV wavelength, UV intensity, additives of reaction media and reactor configuration. Other variables, such as CO_2 pressure, moisture and residence time are also important in photoreducing CO_2 (Wu & Lin, 2005).

Electrochemical Production of Methanol from CO_2 and H_2O: The direct reduction of CO_2 to CH_3OH is known to occur at several types of electrocatalysts including oxidized Cu electrodes. The current stage of the technology is still very experimental. The majority of tests have been performed on a laboratory scale with a purpose of either kinetic analysis or proof-of-concept to examine product distribution for different material and condition combinations (Beck et al., 2010). An advantage of electrochemical CO_2 reduction is that unlike many other hydrocarbon processes it can occur at ambient conditions.

The electrochemical reduction of carbon dioxide to methanol is thermodynamically possible, but there seems to be no well-established technique to achieve this reaction with high current efficiencies close to l00%. Nevertheless, methanol production has been reported with the use of ruthenium: gallium arsenide and RuO_2-TiO_2 mixed cathodes (Le, 2009). Cole and Bocarsly (2010) have reviewed electrochemical reduction processes, including electrochemical CO_2 conversion to methanol. Few studies have investigated the feasibility of this technology, and none has been found to provide an in-depth analysis of its potential industrial implementation.

Applications of Methanol: Methanol has traditionally been used as feed for production of a range of chemicals including acetic acid, formaldehyde and MTBE (Olah, 2009). In recent years, methanol has also been used for other markets such as production of Di-methyl-ether (DME) and olefins by the so-called methanol to olefins process (MTO) or as blendstock for motor fuels. As a liquid fuel, methanol is of interest especially for use in fuel cells (Olah, 2009).

Methanol to Olefins (MTO) explores alternative pathways to produce small olefins, in particular ethylene and propylene. Conventional steam cracker feeds are either natural gas liquids (NGL) or heavy liquids (i.e., naphtha). Ethane cracking, however, is increasing its share as feedstock. A promising alternative route is dehydration of methanol (MTO). Methanol-to-olefins (MTO) was first developed by ExxonMobil (1980s) as part of its methanol-to-gasoline (MTG) process. In the 1990s, UOP and Norsk Hydro built an MTO pilot plant in Norway. Since then, Lurgi has developed its own version of this process, methanol-to-propylene (MTP).

Co_2 to DMC

Dimethyl carbonate (DMC) is a biodegradable and nontoxic chemical acceptable environmentally as a chemical destination of CO_2. It is exempted from VOC classification and can be used as raw material for producing valuable chemicals, including aromatic polycarbonate, and qualifies as an octane booster component in gasoline and diesel.

It is a safer and nontoxic substitute of well-established methylating-carbonylating hazardous chemicals like dimethyl sulfate and phosgene (Souza et al., 2013). Although DMC is presently produced on a relatively small scale, approximately 400 kt/y, its demand has grown strongly in recent times because of its green properties.

Currently, DMC is produced mainly by oxidative carbonylation of methanol (Aoussi et al., 2010). The direct methylation reaction is possible where, according to Ferreira et al. (2013), the most used catalyst is tin, employed as an oxide compound or as an organometallic complex, according to Eq. (31):

$$2\ CH_3OH\ +\ CO_2\ \longrightarrow\ DMC\ +\ H_2O$$

However, direct methylation presents low yields, inferior to 10%, due to the chemical inertness of CO_2 and to the deactivation of catalysts induced by water formation in the reaction (Aouissi, 2010). For large-scale production of DMC from CO_2, one route seems to be promising: the indirect route (IR) for two-step conversion of CO_2 with ethylene oxide (EO) to ethylene carbonate (EC), which then reacts with excess methanol (MeOH) giving DMC and ethylene glycol (EG) as shown in Eqs. (32) and (33).

$$EO\ +\ CO_2\ \longrightarrow\ EC$$

$$EC\ +\ CH_3OH\ \longrightarrow\ DMC\ +\ EG$$

Souza et al. (2013) evaluated IR's performance from technical, economical and environmental standpoints. Accordingly, the authors proposed a process flowsheet with two serial reactors' system followed by an integrated separation section, with extractive distillation using methyl-iso-butyl ketone (IR-MIBK) and ethylene glycol (IR-EG). For environmental performance assessment, Souza et al. (2013) defined Chemical Sequestration of CO_2 (CSC) as the amount of CO_2 consumed by chemical reaction minus the amount of CO_2 emitted by heat, power and purges. Considering a production of 1.3×10^5 t/y of DMC produced, the authors reported CSC values for IR−MIBK and IR−EG of -15.9kt/y and -8.1kt/y, respectively. The negative values of CSC indicate that both alternatives are net emitters and illustrates that fixation of CO_2 in chemical products does not necessarily imply into net CO_2 reduction, and that the main aspect in CO_2 utilization as feedstock is of substituting fossil carbon source for a renewable alternative. Furthermore, Souza et al. (2013) estimated net present values for IR−MIBK and IR−EG of 71.5×10^6 and 106.5×10^6, respectively, and payback times of 5.5 and 4.5, respectively, concluding for economic feasibility of DMC production. It is worth noting that DMC production from CO_2 is already in use at Asahi Chemical Industry.

Biochemical Conversion of Co_2

Biofixation of CO_2 with microalgae is a promising route of utilization of CO_2, as it exhibits fast growth (e.g., Picardo et al., 2013a) and produces numerous high-added-value bioproducts (Grima et al., 2003). Additionally, they do not contain lignin, a fact that renders microalgae better adapted to biochemical valorization. As advantages of CO_2 bioconversion with relation to energy crops, microalgae grow in variable climates on non-arable land with non-potable water, releasing competition with food crops, and are able to use direct flue gases as their carbon source (Fernández et al., 2012). Alternatively, to inject directly flue gases into microalgae cultures, adequate design and operation of the carbonation culture system unit are also necessary, otherwise almost all of the CO_2 fed to the culture would be released into the atmosphere. In this aspect, photobioreactors are more appropriated arrangements. Furthermore, photobioreactors provide cell requirements such as light, temperature, pH, and mixing (Fernández et al, 2012).

Chisti (2007) concluded that microalgae are the only alternative for the sustainable production of biodiesel. Accordingly, the fact that microalgae biomass is rich in lipids and are, hence, high energy density feedstock for fuels and chemicals is of relevance. Picardo et al (2013a) proposed a screening procedure for microalgae selection to meet production objectives such as SYNGAS for the production of synthetic fuels, since the biomass, or the residual biomass obtained after extraction of bioproducts, can be gasified to yield SYNGAS. An attractive alternative in this route is to employ CO_2 as oxidation agent (Butterman and Castaldi, 2007). Butterman and Castaldi (2007) report that the injection of CO_2 and H_2O in gasification increases char reactivity that results in more efficient use of the feedstock with less residual to be post-processed.

According to Grima et al. (2003), production of microalgal biomass can be carried out in fully contained photobioreactors or in open ponds and channels. Biomass productivity depends on species, operational conditions and the choice of ponds (~20g/m².d) or photobioreactor (~50g/m².d) geometry. Open-culture systems are almost always located outdoors and rely on natural light for illumination while closed photobioreactors may be located indoors or outdoors, although outdoor location is more common. Grima et al. (2003) list as biomass harvesting operations centrifugation, filtration or gravity sedimentation, which may be preceded by a flocculation step.

Microalgae contain lipids and fatty acids as membrane components, storage products, metabolites and sources of energy. Microalgae have been found to contain proportionally high levels of lipids (for some species this value can reach 50% oil by weight), with a convenient fatty acids profile and an unsaponifiable fraction allowing a biodiesel production with high oxidation stability (Grima et al., 2013). Lipid accumulation is promoted by stress, notably by nitrogen starvation (Picardo et al., 2013b).

Elemental analysis of carbon content of biomass points to ~50% (Picardo et al., 2013a), what leads to conclude that approximately 2t of CO_2 can be converted into 1t of biomass, potentially amenable to 0.2t of lipids. Its massive extension to the energy sectors constitutes a vast potential for large-volume CO_2 utilization. Fernandez et al. (2012) recognize that microalgae are not a storage strategy because the biomass produced cannot be stored for a long time. Its contribution to reducing CO_2 emissions is only possible if biofuels are produced to replace the fossil fuels use, and allowing the production of other commodities, or by-products from flue gases, which allows one to obtain revenues to mitigate the penalty of carbon capture. To illustrate the potential industrial application of microalgae, Figure 19 shows a schematic of microalgae bioconversion of CO_2 and its downstream processing in a biorefinery arrangement producing long-chain fatty acids (PUFA´s, MUFA´s and PUFA´s), biodiesel, green diesel, gasoline, biogas, urea, N_2 and propylene carbonate.

Monteiro et al. (2010) employed Pareto optimization of what was named an "industrial ecosystem" comprised of a biorefinery of microalgal biomass aiming at maximizing sustainability of the productive arrangement. The authors concluded that increasing the weight of environmental objectives against economic performance might make sectors of the proposed original superstructure of amenable processes unattractive. Therefore, the final structure of a biorefinery of microalgae depends on the priorities set for the productive complex.

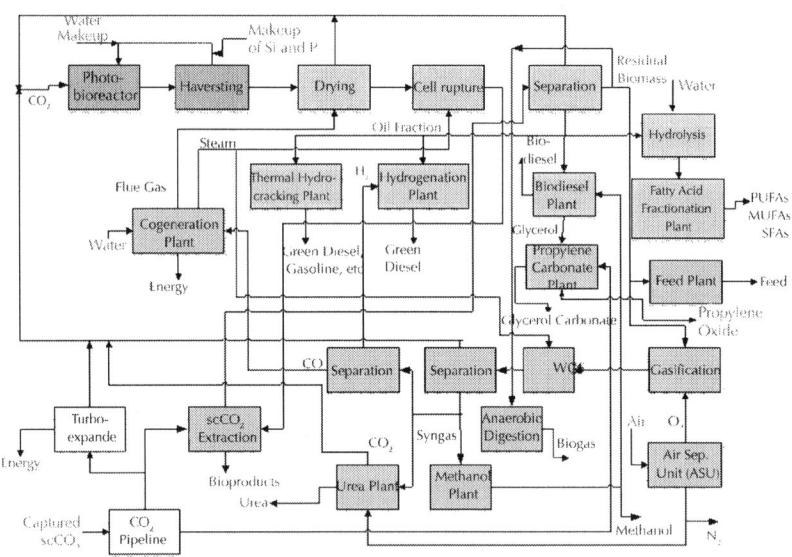

Figure 19. Bioconversion of CO_2 integrated into a biorefinery arrangement. Main external inputs are marked in red.

Some Pilot and Commercial Scale CO_2 Utilization Processes

Polycarbonate: Polycarbonate (PC) is a plastic with impact resistance and heat resistance, mainly produced (4t/y) by reacting CO and Cl_2 to form phosgene as an intermediate material. The phosgene process has a number of disadvantages, including the risk of environmental harm (Ushikubo, 2013). Asahi Kasei Corporation succeeded in commercializing the first non-phosgene polycarbonate production using ethylene oxide and CO_2, a by-product of ethylene oxide synthesis. The Asahi Kasei Process has as co-products high-purity monoethylene glycol (MEG). The process employs reactive distillation in the monomer production and gravity-utilized, non-agitation polymerization reactor in the melt polymerization. The monomer process consists of 3 production steps, ethylene carbonate (EC) from CO_2 and EO, dimethyl carbonate (DMC) and MEG from EC and MeOH, and diphenyl carbonate (DPC) and MeOH from DMC and phenol (PhOH). All intermediates are recycled. The by-produced PhOH is recycled to the monomer process (Sinshuke et al., 2010). According to Ushikubo (2013), the new process reduces CO_2 emissions by 0.173kg per one kg of polycarbonate. Five commercial plants using the Asahi Kasei process are operating in Taiwan (150,000 t/y), Korea (2 plants of 65,000 t/y), Russia (65,000 t/y) and Saudi Arabia (260,000 t/y) (Shinsuke et al., 2010).

Monoethylene Glycol (MEG): MEG is used as an antifreeze and as a raw material for the production of polyester fibers and resins, mainly PET (Ushikubo, 2013). An expanding market share is foreseen in natural gas industry, where MEG is added into the pipeline or the gas conditioning process, either as hydrate inhibitors or for dehydration purposes to protect downstream pipelines. The pipeline can extend for thousands of kilometers and MEG is injected to inhibit hydrate formation, avoiding plugging (e.g.,Statoil's Snøhvit field, Pettersen, 2011). Concerning the production process, a new technology was developed by Mitsubishi Chemical Corporation with 99% selectivity while the conventional (non-catalytic) process has selectivity around 89% (Kawabe, 2010). The conventional technology produces as co-product DEG and TEG (di- and triethylene glycols, whose demand is expanding at only 2-3% as opposed to MEG expansion (world demand amounts to 17t/y). Mitsubishi technology uses a two-step catalytic synthesis: production of ethylene carbonate (EC) as intermediate followed by EC hydrolysis under almost stoichiometric condition, while the conventional hydrolysis occurs at a higher H_2O/ethylene oxide molar ratio, according to Eqs. (34) and (35).

$$EO + CO_2 \longrightarrow EC \text{ (carbonation)}$$

$$EO + H_2O \longrightarrow EG + CO_2 \text{ (hydrolysis)}$$

The product purification is simpler (water removal and MEG purification distillation columns, while the conventional process has 4 distillation columns: water, MEG, DEG and TEG columns). Furthermore, CO_2 remains in closed loop. Ushikubo (2013) reports that several commercial plants operating with the new (and greener) technology save resources and energy and reduces the amount of wastewater and CO_2 production. It is worth noting that MEG is a co-product of the production of DMC in a process where Eq. (4.24) is replaced by Eq. (40).

Polyurethane: Bayer (2013) targets the production of polyurethane, via the utilization of CO_2 as feedstock. In this route, CO_2 is converted to polyols (HO-R-OH) which reacts with isocyanate to yield polyurethane. The conversion of CO_2 starts with its reaction with an epoxide, (propylene oxide) of higher energy content, in a catalytic route. Polyol with 30%CO_2 had 2.64 kg of equivalent CO_2 emissions. A maximum theoretical value of 43% of CO_2 can be incorporated in the polyol (Bayer, 2010). The development started in 1669, ended its laboratory scale in 2009, and is moving to industrial implementation of the named "Dream Production" with a pilot plant in Leverkusen to produce polyol, for testing purposes. In early 2013, the new method was successfully converted from the production of discrete quantities to continuous production, a key intermediate step for the industrial-scale production of CO_2-based polyurethane, which Bayer is targeting for 2015.

Emerging Co_2 Utilization Processes

The fixation of CO_2 into chemicals and polymers will not substantially contribute to a reduction in antropogenic GHG emissions given the current energy demand. Nevertheless, using CO_2 as a feedstock meets the requirements of sustainable development. An insight into advanced process concepts focus on chemical sequestration of CO_2 creating manufactured products from captured CO_2 with large potential markets. Integration of capture technologies into energy production schemes or oil and gas refining installations is the idea behind cPSE approach, namely, to abate chemical emissions while producing industrial products.

Formic acid: According to Armstrong et al. (2013), the amount of energy required to utilize carbon dioxide as a feedstock largely depends on the oxidation state of the intended products. The next-highest oxidation state molecules from CO_2 are formic acid (HCOOH) and carbon monoxide (CO). So, carbon dioxide utilization to manufacture formic and other carboxylic acids is a relatively low-energy transformation. Formic acid has numerous applications, including food technology, agriculture, and the leather and rubber industries. Moreover, it has recently been considered as a promising candidate material for hydrogen storage and it is an important chemical with numerous applications. Moreover, formic acid has limited uses for further conversion, except reduction to methanol. The industrial methods used for its production employ CO as a raw material. Maihom et al. (2013) concluded that a first step occurs where CO_2 is hydrogenated to a formate intermediate. In the second step, the formate is further hydrogenated into formic acid. The hydrogenation of CO_2 would complete the chemical loop for hydrogen storage using CO_2. The complementary step is the catalyzed decomposition of formic acid to pure H_2 and reusable CO_2.

Carbon dioxide-based copolymers: The synthesis of organic carbonates has been one of the most widely studied areas of CDU. Typically, CO_2 is inserted into a molecule without the loss of any atoms in either the co-reactant or the gas itself (Armstrong, 2013). Carbonates are formed by the insertion of a CO_2 molecule into a guest co-reactant, typically an epoxide. Poly(propylene carbonate) (PPC), an alternating copolymer of CO_2 and propylene oxide, is one of the emerging low-cost biodegradable plastics. The fast development in catalyst design and performance improvement for PPC has created new chances for the chemical industry. In particular, high molecular weight PPC from rare earth ternary catalyst is becoming an economically viable biodegradable plastic with tens of thousands of tons produced per year, providing a new solution to overcoming the problem of high cost in biodegradable plastics (Qin and Wang, 2010). According to Qin and Wang (2010), with the continuous improvement in catalyst systems, commercialization of CO_2 copolymer is possible. The authors report industrial activities by Empower Materials producing polypropylene carbonate (QPAC®40), polyethylene carbonate (QPAC®25), polybutylene carbonate (QPAC®60), and polycyclohexene oxide(QPAC®130) on a pilot scale.

Electrochemical Reduction of CO_2: Delacourt (2010) studied the electrochemical conversion of CO_2 into SYNGAS. The driver of the proposed route is that renewable energies (e.g., solar and wind) are only alternatives to fossil fuel as they are not available on demand, thus requiring storage. Delacourt (2010) lists as a storage opportunity the conversion to liquid fuels (e.g., methanol), in which SYNGAS is the required intermediate, by converting solar energy into electricity through photovoltaic arrays, and then by using this electricity to produce fuels by electrolysis. Evolved H_2 reacts with CO_2 in a water-gas-shift reactor to make CO (and H_2O). The resulting SYNGAS is converted to methanol. Delacourt (2010) decided for a low-temperature technology (room temperature) although reported that high-temperature electrolysis (800 to 900°C) could be an attractive alternative. Because of the relatively low solubility of CO_2 in water under ambient conditions, gas-diffusion electrodes were applied to operate at higher current densities, and ion-exchange membrane was used as the electrolytic medium to limit gas crossover resulting in a decrease of the current efficiency of the electrochemical cell. Catalysts capable of reduction of CO_2 to CO at low overpotentials were selected.

Light-Driven Technologies: The rubisco enzyme is probably the most abundant enzyme of the biosphere. The fixation of CO_2 and its transfer to organic substrates in the Calvin cycle leads by way of starch to an annual production of 10^{11} t of biomass (Walther et al., 1999). With the development of catalysts able to reproduce the key steps of photosynthesis, water and sunlight would ultimately be the only needed sources for clean energy production. Light driven technologies under development include (a) photoelectrochemical cells where CO_2 present in a moistened gas stream is converted into organic molecules based on the photooxidation of water into oxygen gas O_2, protons H_+, and electrons. The conversion of CO_2 occurs at the photocathode and involves the generated protons, electrons and the "fuel" CO_2 (Kayaert et al., 2013); (b) direct water oxidation - photocatalytic water splitting - to produce H_2 and O_2 over a metal-oxide-based photocatalyst using solar energy (Maeda and Domen, 2013); (c) hydrogen-producing systems consisting of a hydrogen-evolving catalyst linked to a photosensitizer (Badura et al., 2012). Although promising alternatives, biomimetic CO_2 conversions are still in its early stage of technological development.

CO_2 Mineralization for Environmental Remediation: Lim et al. (2013) reviewed the application of carbonation to solidify or stabilize solid combustion residues from municipal solid wastes, paper mill wastes, etc. and contaminated soils, and to manufacture precipitated calcium carbonate. For instance, the red mud - a highly alkaline waste of Bayer's process - can be treated by absorption of CO_2. Machado (2012) analyzed the process of red mud carbonation with the exhausted gases from the alumina production calcinators, by developing a dynamic model representative of the mass and energy balances involved in the process, and chemical reactions occurring in the mud under carbonation. Machado (2012) was able to predict the species behavior, as well as the decrease in mud pH and the rebound phenomenon observed when the CO_2 concentration is reduced. The transient profile of the main process responses indicated a substantial reduction of CO_2 concentration in the output gas, in consequence of tons of CO_2 captured, and a significant reduction in mud pH. Concerning other environmental applications, Lim et al. (2013) report that carbonated products can be utilized as aggregates in the concrete industry and as alkaline fillers in the paper (or recycled paper) industry. Mineral carbonation consist in reacting CO_2 and Ca or Mg-bound compounds such as wollastonite ($CaSiO3$), olivine ($Mg2SiO4$), and serpentine ($Mg_3Si2O_5(OH)_4$). As a result, CO_2 is stably stored in final products such as $CaCO_3$ and $MgCO_3$. Last, the accelerated carbonation of solid wastes containing alkaline minerals such as Ca and Mg before their landfill treatment is effective for decreasing the mobility of heavy metals by adjusting pH to below 9.5 at which their solubility is lowest.

Non-Conversion Utilization of Co_2

CO_2 utilization that does not involve its chemical conversion is an alternative destination of captured emissions. Among such alternatives the injection of supercritical CO_2 into depleted oil wells to enhance the further recovery of oil is well established. Indeed, this is presently the only commercially viable technology adding value to large volumes to CO_2 in the order of magnitude of emissions from fossil fuel based energy generation. It has been estimated that CO_2 injection can enhance oil recovery from a depleting well by about 10 to 20 % of the original oil in place. Similarly, CO_2 can be used to recover methane from unmined coal seams. It has been estimated that, in the U.S. alone, 89 billion barrels of oil could technically be recovered using CO_2, leading to a storage of 16 Gt of CO_2 in the depleted oil reservoirs (DNV, 2011).

The use of supercritical CO_2 as a solvent in processing chemicals (e.g., flavor extraction) is also well established. New uses of supercritical CO_2 in chemical processing are emerging, and have the added benefit of reducing water usage. Supercritical CO_2 is also being explored as a heat transfer fluid for some geothermal applications. These non-conversion methods of utilization constitute a significant fraction of the total CO_2 emissions (DNV, 2011).

Enhanced Oil Recovery (CO_2-EOR): Through CO_2-EOR, oil producers inject CO_2 into wells to help sustain production in otherwise declining oil fields. The main goal of this technology is to draw more oil to the surface. In 2012, CO_2-EOR accounted for 6% of current U.S. domestic oil production. The limited CO_2 source is the main barrier to reaching higher levels of CO_2-EOR production due to insufficient supplies of affordable CO_2. With the discovery of offshore gas fields with high CO_2 contents in Brazil, there is a great opportunity to implement CO_2-EOR at those fields.

Furthermore, the offshore removal of acid gases poses a choice of onshore processing against offshore processing. Factors like safety and operability may favor onshore processing in comparison with offshore processing. The proper on land disposal of the CO_2 removed from natural gas requires the construction of CO_2 pipelines to transport CO_2 to offshore EOR applications. Another aspect is the high cost of ship hulls as shifting CO_2 removal to onshore facilities releases the weight shipped, which could overload the cost of building the required CO_2 pipelines.

In the option of onshore processing, CO_2 rich natural gas would be available as feedstock to SYNGAS production from CO_2 reforming, besides CO_2 separation and transport back to oil fields for CO_2-EOR. The current estimated cost gap for CGS from power, steel and cement plants is several times larger than the current CO_2 market price, and downward pressure on this market price is likely to increase. Investments in CO_2 reuse technologies need to be assessed as a screening procedure among potential alternatives.

CONCLUDING REMARKS

Technologies for utilization of CO_2 amenable to commercial scales are presently a very small fraction of anthropogenic CO_2 emissions, and very endothermic due to the inertness of CO_2, what reduces their abatement potential. Furthermore, chemical and biochemical conversion of CO_2 presents a sequestration potential that is orders of magnitude lower than the CO_2 emissions associated to energy generation from fossil fuels.

Geographical synergies of CO_2 supply (power plant emissions or natural gas processing) should guide in the medium term feasible utilization alternatives. The main synergy is identified in offshore gas processing and EOR, which, due to the economic benefit, process scale and maturity, stands as the most relevant utilization route in the short to medium term. Furthermore, most of the emerging alternatives reviewed are at their early stage of technological development.

However, CO_2 stands as a promising renewable feedstock to the chemical industry, which has been limited to oil, natural gas, coal and, recently, biomass. Such as posed, SYNGAS based conversions to the downstream supply chain is a route for flexibility of raw materials. Gasification of a variety of feedstock can lead to SYNGAS. Furthermore, expanding non-conventional gas supply enforces natural gas reforming in the upstream of the chemical supply chain. CO_2 captured from emissions and natural gas processing may drop into the supply chain via Dry Reform. As SYNGAS derived products, hydrogen, methanol and synthetic fuels (e.g., olefins, naphta, diesel, lubricants and kerosene) from Fischer-Tropsch process are likely to dominate the scenario.

Additionally, methanol (MeOH) is expected to grow in relevance either as hydrogen carrier and as intermediate product such as feedstock to MeO (Methanol to Olefins) process, as well as trans-esterification agent in biodiesel and dimethyl carbonate (DMC) production processes. Nowadays, the interest in DMC has grown significantly because it is considered to be a safe and nontoxic substitute for well-established methylating and carbonylating agents (e.g., phosgene), and has potential as an oxygen-containing fuel additive. There are several technological routes to produce DMC, however, the one route considered promising for large-scale commercialization is the trans-esterification of ethylene carbonate (EC) with methanol. In this indirect route, EC is obtained by a previous reaction of CO_2 with ethylene oxide. The route yields DMC and ethylene glycol (EG) as co-products in equimolar ratio.

The use of CO_2 as a carbon source in the synthesis of chemicals, in contrast to disposal, reduces dependence on fossil fuels, generates profit and is in line with a sustainable chemical industry. However, the actual use of CO_2 corresponds to about 0.4% of the potential CO_2 suitable to be converted to chemicals (Navarro et al., 2013).

Finally, large-scale utilization of CO_2 require energy efficient CO_2 capture technologies and an expansion of CO_2 transportation infrastructure.

ACKNOWLEDGEMENTS

O. Araujo and J.L. Medeiros kindly acknowledge CNPq for scholarships and financial grants; and CAPES for grant no. 113/2008.

REFERENCES

1. ADEME (Agence de l'Environnement et de la Maîtrise de l'Energie), Panorama des voies de valorisation du CO2 http://www2.ademe.fr/servlet/getDoc?cid=96&m=3&id= 72052&p1=30&ref=12441 (2010).
2. Aouissi, A., Al-Othman, Z.A., Al-Amro, A. Gas-Phase Synthesis of Dimethyl Carbo- nate from Methanol and Carbon Dioxide over Co1.5PW12O40 Keggin-Type Hetero- polyanion. Int. J. Mol. Sci., 11, 1343-1351 (2010).
3. Aresta, M. Perspectives in the use of carbon dioxide. Quím. Nova, 22 (2), (1999). http://dx.doi.org/10.1590/S0100-40421999000200019
4. Aresta, M., Aresta, Michele. Carbon Dioxide as Chemical Feedstock, WILEY-VCH Verlag GmbH & Co., (2010).

5. Armstrong, Katy, Dawson, George and Styring, Peter. Recent Advances in Catalysis for Carbon Dioxide Utilization. The Catalyst Review, vol. 26, issue 2, 6-13, (2012).
6. Ashcroft, A.T., Cheetham, A.K., Green, M.L.H., Vernon, P.D.F, Partial oxidation of methane to synthesis gas using carbon dioxide Nature 352, 225 - 226 (18 July 1991).
7. Badura, A., Guschin, D., Esper, B., Kothe, T., Neugebauer, S., Schuhmann, W., Rögn- er, M. Photo-Induced Electron Transfer Between Photosystem 2 via Cross-linked Re- dox Hydrogels. Electroanalysis 20(10), 1043–1047 (2012).
8. Bayer. Use of carbon dioxide for the production of plastics. http://www.materials- cience. bayer.com/~/media/Bms/Documents/Publications/CO2-Pro- jekte_kurz_EN.ashx. Accessed on 11/27/2013 (2013).
9. Bayer. A Dream Comes True http://solutions.bayertechnology.com/fileadmin/ user_upload/sat_pages/technologyImpulse/articles/A-Dream-Comes-True.pdf. Ac- cessed on 11/27/2013 (2010).
10. Beck, J., R. Johnson, R., Naya, T., Electrochemical Conversion of Carbon Dioxide to Hydrocarbon Fuels, EME 580, Spring (2010).
11. Bradford, M. C. J., Vannice, M. A., CO2 Reforming of CH4, Catalysis Reviews, Vol. 41(1), 1-42, (1999).
12. Bustamante, F., Enick, R. Rothenberger,K., Howard, B., Cugini, A., Ciocco, M. Mor- reale, B., Kinetic Study of the Reverse Water Gas Shift Reaction in High-Tempera- ture, High Pressure Homogeneous Systems, Fuel Chemistry Division Preprints, 47(2), 663, (2002).
13. Butterman, H.C., Castaldi, M.J. Influence of CO2 Injection on Biomass Gasification. Ind. Eng. Chem. Res., 46, 8875-8886 (2007). CO2 Utilization: A Process Systems Engineering Vision http://dx.doi.org/10.5772/57560 81
14. Byron, S.R.J., Loganthan, M., Shantha, M.S., A Review of the Water Gas Shift Reac- tion. International Journal of Chemical Reactor Engineering 8, 1–32, (2010).
15. Chisti, Y. Biodiesel from microalgae. Biotechnol Adv, 25, 294–306 (2007).

16. Cole, E.B., Bocarsly, A.B., Photochemical, Electrochemical, and photoelectrochemical Reduction of Carbon Dioxide, in "Carbon Dioxide as Chemical Feedstock", M. Are- sta, Editor, Wiley-VCH Verlag GmbH & Co.: Weinheim. (2010).
17. de Medeiros, J.L., Versiani, B., Araújo, O.Q.F. A model for pipeline transportation of supercritical CO2 for geological storage. The Journal of Pipeline Engineering, 4th Quarter, 253-279 (2008).
18. de Medeiros, J.L., Nakao, A., Grava, W.M., Nascimento, J.F., Araújo, O.Q.F. Simula- tion of an Offshore Natural Gas Purification Process for CO2 Removal with Gas Liq- uid Contactors Employing Aqueous Solutions of Ethanolamines. Industrial & Engineering Chemistry Research, 52, 7074-7089 (2013a).
19. de Medeiros, J.L., Barbosa, L.C., Araújo, O.Q.F. An Equilibrium Approach for CO2 and H2S Absorption with Aqueous Solutions of Alkanolamines: Theory and Parame- ter Estimation. Industrial & Engineering Chemistry Research, 52, 9203-9226 (2013b).
20. Delacourt, C. Electrochemical reduction of carbon dioxide and water to syngas (CO + H2) at room temperature. Available at http://charles.delacourt.free.fr/Postdoc-manu- script_Charles-Delacourt.pdf (2010).
21. DNV. Carbon Dioxide Utilization Electrochemical Conversion of CO2 – Opportuni- ties and Challenges. Available online at http://www.dnv.com/binaries/DNV-posi- tion_paper_CO2_Utilization_tcm4-445820.pdf (2011).
22. Edwards, J.H., Potential sources of CO2 and the options for its large-scale utilisation now and in the future. Catalysis Today, 23, 59-66, (1995).
23. Ferreira, H.B.P., Vale, D.L., Mota, C.J.A., Miranda, J.L. Experimental Design for CO2 Conversion into Dimethylcarbonate (DMC) using Bu2SnO at Subcritical Conditions. Brazilian Journal of Petroleum and Gas, 6(3), 93-104, (2012).
24. Fernández, F.G.A., González-López, C.V., Sevilla, J.M.F., Grima, E.M. Conversion of CO2 into Biomass by Microalgae: How Realistic a Contribution May it be to Signifi- cant CO2 Removal? Appl Microbiol Biotechnol, 96, 577–586 (2012).
25. Gangadharan, P., et al., Evaluation of the economic and environmental impact of combining dry reforming with steam

reforming of methane. Chem. Eng. Res. Des. (2012), http://dx.doi.org/10.1016/j.cherd.2012.04.008

26. Genesis. Equation of State Prediction of Carbon Dioxide Properties. Project King- snorth Carbon Capture & Storage Project. CP-GNS-FAS-DRP-0001. http://www.decc.gov.uk/assets/decc/11/ccs/chapter6/6.23-equation-of-state-prediction-of-carbondioxide-properties.pdf, (2011). 82 CO2 Sequestration and Valorization

27. Ginsburg, J. M.; Pina, J.; El Solh, T.; de Lasa, H. I.Coke formation over a nickel cata- lyst under methane dry reforming conditions: Thermodynamic and kinetic models Ind. Eng. Chem. Res., 44 (14) 4846– 4854, (2005).

28. Grima, E.M., Belarbia, E.-H., Acién Fernández, F.G., Medina, A.R., Chisti, Y. Recov- ery of microalgal Biomass and Metabolites: Process Options and Economics. Biotech- nology Advances, 20, 491–515 (2003).

29. Gunardson, H. Industrial Gases in Petrochemical Processing; Marcel Dekker: New York, (1998).

30. Hartley, M., Tam, I., Non-Sequestration Utilization Options for Carbon Dioxide ($CO2$), Nexant´s ChemSystems PERP09/10S10, (2012).

31. Ho, M.T., Wiley, D. E. Economic Evaluation of Membrane Systems for Large Scale Capture and Storage of $CO2$ Mixtures. Report, UNESCO Centre for Membrane Sci- ence, The University of New South Wales (2005).

32. Hu, B., Guild, C., Suib, S.L., Thermal, electrochemical, and photochemical conversion of $CO2$ to fuels and value-added products, Journal of $CO2$ Utilization, 1, 18–27, (2013).

33. IEA. Technology Roadmap - Energy and GHG Reductions in the Chemical Industry via Catalytic Processes. Technology Perspectives. Available online at: http://iea.org/ media/freepublications/technologyroadmaps/TechnologyRoadmapCatalyticProces-sesAnnexes.pdf, (2013).

34. Jarungthammachote, S., Combined Partial Oxidation and Carbon Dioxide Reforming Process: A Thermodynamic Study, American Journal of Applied Sciences 8 (1): 9-14, (2011).

35. Jiang, Z.; Liao, X.; Zhao, Y., Comparative study of the dry reforming of methane on fluidized aerogel and xerogel Ni/Al2O3 catalysts. Applied Petrochemical Research, p. 1-9, (2013).
36. Kahle, L. C. S., Roussière, T., Maier, L., Delgado, K.H., Wasserschaff, G., Schunk, S.A., Deutschmann, O., Methane Dry Reforming at High Temperature and Elevated Pressure: Impact of Gas-Phase Reactions, Industrial & Engineering Chemistry Research, 52 (34), (2013).
37. Kaiser, P., Unde, R.B., Kern, C., Jess, A. Production of Liquid Hydrocarbons with CO2 as Carbon Source based on Reverse Water-Gas Shift and Fischer-Tropsch Synthesis. Chemie Ingenieur Technik. Special Issue: Reaktionstechnik. 85(4), 489–499, (2013).
38. Kawabe, K. Development of Highly Selective Process for Mono-Ethylene Glycol Pro- duction from Ethylene Oxide via Ethylene Carbonate Using Phosphonium Salt Cata- lyst. Catal Surv Asia 14, 111–115, (2010).
39. Kayaert, S., Martens, J., Masschaele, K. Photo-Electrochemical Cell. United States Pat- ent Application 20130026029, (2013). CO2 Utilization: A Process Systems Engineering Vision http://dx.doi.org/10.5772/57560 83
40. King, D. The Future of Industrial Biorefineries. World Economic Forum, (2010).
41. Kuwabata, S., Nishida, K., Tsuda, R., Inoue, H., Yoneyama, H., Photochemical Re- duction of Carbon Dioxide to Methanol Using ZnS Microcrystallite as a Photocatalyst in the Presence of Methanol Dehydrogenase, J. Electrochem. Soc. 141(6), 1498, (1994).
42. Kuwabata, S., Tsuda, R., Yoneyama, H., Electrochemical Conversion of Carbon Diox- ide to Methanol with the Assistance of Formate Dehydrogenase and Methanol Dehy- drogenase as Biocatalysts. J. Am. Chem. SoC., 116, 5431-5443, (1994).
43. Kurz, G. , Teuner, S., Calcor process for carbon monoxide production, Erdoel & Kohle, Erdgas, Petrochemie, 43(5), 171-172, (1990).
44. Lachowska, M., Skrzypek, J., Ga, Mn, and Mg Promoted Copper/Zinc/Zirconia-Cata- lysts for Hydrogenation of Carbon Dioxide to Methanol, Carbon Dioxide Utilization for Global Sustainability, Elsevier, Amsterdam, (2004).

45. Le, M.T.H., Electrochemical Reduction of CO2 to Methanol, MSc Thesis. Graduate Faculty of the Louisiana State University and Agricultural and Mechanical College, (2011).
46. Lee, D., Lee, J.-Y., Lee, J.S., Effects of Palladium Particle Size in Hydrogenation of Carbon Dioxide to Methanol over Pd/ZnO Catalysts, Carbon Dioxide Utilization for Global Sustainability, Elsevier, Amsterdam, (2004).
47. Levitan, R., Levy, M., Rosin, H., Rubin, R., Closed-loop operation of a solar chemical heat pipe at the Weizmann Institute solar furnace. Solar Energy Materials vol. 24 is- sue 1-4, 464-477, (1991).
48. Levy, M., Levitan, R., Rosin, H., Rubin, R., Solar energy storage via a closed-loop chemical heat pipe. Solar Energy, vol. 50, issue 2, 179-189, (1993).
49. Li, H.. Thermodynamic Properties of CO2 Mixtures and Their Applications in Ad- vanced Power Cycles with CO2 Capture Processes. Energy Processes Department of Chemical Engineering and Technology Royal Institute of Technology, Stockholm, Sweden. TRITA-CHE Report 2008:58, (2008).
50. Li, H., Yan, J.. Evaluating Cubic Equations of State for Calculation of Vapor–Liquid Equilibrium of CO2 and CO2 -Mixtures for CO2 Capture and Storage Processes. Chem. Eng. and Technology/ Energy Process, Royal Institute of Technology, (2009).
51. Lim, M., Han, G.C., Ahn, J.W., You, K.S. Environmental Remediation and Conver- sion of Carbon Dioxide (CO2) into Useful Green Products by Accelerated Carbona- tion Technology. Int. J. Environ. Res. Public Health, 7, 203-228, (2010).
52. Li, M.W., Xu, G.H., Tian, Y.L., Chen, L., Fu, H.F. Carbon dioxide reforming of meth- ane using DC corona discharge plasma reaction. J. Physical Chem. A., 108: 1687-1693, (2004).
53. Machado, R.B.P. Avaliação Técnica de Processo de Carbonatação de Lama Vermelha com Gás Exausto de Calcinadores de Alumina. MSc Thesis. Graduate Program in 84 CO2 Sequestration and Valorization Chemical and Biochemical Process Technology, School of Chemistry, Federal Univer- sity of Rio de Janeiro, (2012).
54. Maeda, K. Domen, K. 7.22 – Photochemical Water Splitting Using Nanostructured Metal Oxides. Reference Module in Chemistry,

Molecular Sciences and Chemical En- gineering Comprehensive Inorganic Chemistry II (Second Edition). From Elements to Applications, 587–614. Vol. 7: Surface Inorganic Chemistry and Heterogeneous Catal- ysis, (2013).

55. Maiohm, T., Wannakao, S., Boekfa, B., Limtrakul, J. Production of Formic Acid via Hydrogenation of CO_2 over a Copper-Alkoxide-Functionalized MOF: A Mechanistic StudyJ. Phys. Chem. C, 117, 17650−17658, (2013).

56. McCoy, S.T. The economics of CO_2 transport by pipeline and storage in saline aqui- fers and oil reservoirs. PhD Thesis, Carnegie-Mellon University, Pittsburgh, USA, (2008).

57. Meunier, F.C., Tibiletti, D., Goguet, A., Shekhtman, S., Hardacre, C., Burch, R., On the complexity of the water-gas shift reaction mechanism over a Pt/CeO_2 catalyst: Ef- fect of the temperature on the reactivity of formate surface species studied by oper- ando DRIFT during isotopic transient at chemical steady-state, Catalys Today, 126, 143, (2007).

58. Moon, D.J., Ryu, J.W., Kang, D.M., Lee, B.J., Ahn, B.S., CO_2 Reforming by CH_4 over Ni-YSZ Modified Catalysts in Carbon Dioxide Utilization for Global Sustainability, Elsevier, Amsterdam, (2004).

59. Monteiro, J.G.M.S., Silva, P.A., Araújo, O.Q.F., de Medeiros, J.L. Pareto Optimization of an Industrial Ecosystem: sustainability maximization. Brazilian Journal of Chemi- cal Engineering, 27, 429-440, (2010).

60. Nakao, A., Macedo, A.P.F., Versiani, B.M., Araújo, O.Q.F., and de Medeiros, J.L., Modeling of Flowcharts of Permeation through Membranes for Removal of CO_2 of Natural Gas, 10th International Symposium of Process System Engineering, PSE-2009, Computer Aided Chemical Engineering, 27, 1875-1880, ISBN 978-0-444-53472-9, Elsevier, (2009)

61. Navarro, R., Pawelec, B., Alvarez-Galván, M. C., Guil-Lopez, R. Al-Sayari, S., Fierro, J. L. G. Renewable SYNGAS Production via Dry Reforming of Methane, in M. de Fal- co et al., CO_2: A Valuable Source of Carbon Green Energy and Technology, pp 45-66, Spring-Verlag London, (2013).

62. Nagaoka, K., Seshan, K., Aika, K., Lercher, J.A., Carbon deposition during carbon di- oxide reforming of methane-comparison between Pt/Al_2O_3 and Pt/ZrO_2. J. Catalysis, 197: 34-42, (2001).

63. O'Connor A.M., Ross J.R.H. The Effect of O2 Addition on the Carbon Dioxide Re- forming of Methane over Pt/ZrO2 . Catalysts. Catal. Today, 46 (2-3), 203–210, (1998). CO2 Utilization: A Process Systems Engineering Vision http://dx.doi.org/10.5772/57560 85
64. Oi, L.E. CO2 removal by absorption: challenges in modelling. Mathematical and Computer Modelling of Dynamical Systems, 16(6), 511–533 (2010).
65. Olah, G.A., Goeppert, A., Prakash, G.K.S., Beyond Oil and Gas: The Methanol Econo- my, 2nd ed., Wiley VCH, Weinheim, Germany, (2009).
66. Olah, G.A., Goeppert, A., Prakash, Czaun, M., G.K.S., Bi-reforming of Methane from Any Source with Steam and Carbon Dioxide Exclusively to Metgas (CO–2H2) for Methanol and Hydrocarbon Synthesis, J. Am. Chem. Soc., 135 (2), 648–650, (2013).
67. Ormerod, W., Riemer, P., Smith, A. Carbon Dioxide Utilisation. IEA Greenhouse Gas R&D Programme, http://www.ieagreen.org.uk/sr4p.htm, (1995).
68. Pettersen, J. Snøhvit field development. Available online at http://www.ipt.ntnu.no/ ~jsg/ undervisning/prosessering/gjester/LysarkPettersen2011.pdf, (2011).
69. Petrobras. Visão Geral da Petrobras. Available online at: http://www.investidorpetro- bras.com.br/pt/apresentacoes/visao-geral-da-petrobras.htm
70. Pettersen, J., Snøhvit field development, TEP4520, Statoil, (2011).
71. Picardo, M.C., de Medeiros, J.L., Monteiro, J.G.M., Chaloub, R.M., Giordano, M., Ofé- lia, Q.F. A methodology for screening of microalgae as a decision making tool for en- ergy and green chemical process applications. Clean Technologies and Environmental Policy, 15, 275-291, (2013a).
72. Picardo, M.C., de Medeiros, J.L., Ofélia, Q.F., Chaloub, R.M. Effects of CO2 Enrich- ment and Nutrients Supply Intermittency on Batch Cultures of Isochrysis galbana. Bioresource Technology, 143, 242-250, (2013b).
73. Qin, Y., Wang, X. Carbon dioxide-based copolymers: Environmental benefits of PPC, an industrially viable catalyst. Biotechnol. J., 5, 1164–1180, (2010).

74. Raju, A.S.K., Park, C.S., Norbeck, J.M., Synthesis gas production using steam hydro- gasification and steam reforming. Fuel Process. Technol. 90, 330–336, (2009).
75. Richardson, J.T., Paripatyadar, S.A. , Carbon dioxide reforming of methane with sup- ported rhodium Appl. Catal. 61, 293-309, (1990).
76. Rostrup-Nielsen, J., 40 years in catalysis, Catalysis Today, 111, 4-11, (2006).
77. Rostrup-Nielsen, J., Christiansen, L.J. Concepts in syngas Manufacture. Catalytic Ser- ies, V. 10. World Scientific, 392 pp, 2011. ISBN: 978-1-84816-567-0, (2011).
78. Saito, M., Murata, K., Development of high performance Cu/ZnO-based catalysts for methanol synthesis and the water-gas shift reaction, Catalysis Surveys from Asia, 8, 285-294, (2004).
79. Sakakura, T., Choi, J., Yasuda, H. Transformation of carbon dioxide. Chem Rev, 107, 2365-2387, (2007). 86 CO2 Sequestration and Valorization
80. Shinsuke, F., Isaburo, F., Masahiro, T., Kazuhiro, O., Hiroshi, H., Muneaki, A., Kazu- mi, H., Kyosuke, K. A Novel Non-Phosgene Process for Polycarbonate Production from CO2 : Green and Sustainable Chemistry in Practice. Catalysis surveys from Asia, 14(3-4), 146-163, (2010).
81. Song, C., Pan, W., Tri-Reforming of Methane: A Novel Concept for Synthesis of In- dustrially Useful Synthesis Gas with Desired H2 /CO Ratios Using CO2 in Flue Gas of Power Plants without CO2 Separation, Prepr. Pap.-Am. Chem. Soc., Div. Fuel Chem., 49 (1), 128, (2004).
82. Song, C. Global Challenges and Strategies for Control, Conversion and Utilization of CO2 for Sustainable Development Involving Energy, Catalysis, Adsorption and Chemical Processing. Catalysis Today, 115, 2–32, (2006).
83. Souza, M.M.V.M., Schmal, M., Methane conversion to synthesis gas by partial oxida- tion and CO2 reforming over supported platinum catalysts, Catalysis Letters, 91(1–2), 11-17, (2003).
84. Souza, L.F.S., Ferreira, P.R.R., de Medeiros, J.L.M., Alves, R.M.B., Araújo, O.Q.F. Pro- duction of DMC from CO2 via Indirect Route: Technical−Economical−Environmental Assessment and Analysis Sustainable Chem. Eng., dx.doi.org/10.1021/sc400279n, (2013).

85. Span, R., Wagner, W. A New Equation of State for Carbon Dioxide Covering the Flu- id Region from the Triple-Point Temperature to 1100 K at Pressures up to 800 MPa. J. Phys. Chem. Ref. Data, 25(6), (1996).
86. Spath, P.L., Dayton, D.C., Preliminary Screening – Technical and Economic Assess- ment of Synthesis Gas to Fuels and Chemicals with Emphasis on the Potential for Bi- omass-Derived SYNGAS. National Renewable Energy Lab Golden Co., (2003).
87. Stoczynski, J., Grabowski, R., Koslowska, A., Lachowska, M., Skrzypek, J., Effect of Additives and a Preparation Method on Catalytic Activity of $Cu/ZnO/ZrO2$ System in the Carbon Dioxide Hydrogenation to Methanol, Carbon Dioxide Utilization for Global Sustainability, Elsevier, Amsterdam, (2004).
88. Tanaka, Y., Utaka, T., Kikuchi, R., Sasaki, K., Eguchi, K., CO removal from reformed fuel over $Cu/ZnO/Al2O3$ catalysts prepared by impregnation and coprecipitation methods, Applied Catalysis A: General, 238, 11–18, (2003).
89. Teuner, S., Make carbon monoxide from carbon dioxide, Hydrocarbon Processing, International Edition, 64(5), 106-7, (1985).
90. Teuner, S. et al., The Calcor Standard and Calcor Economy Processes, in Oil and Gas European Magazine, 44-46, (2001).
91. Treacy, D., Ross, J.R.H. The Potential of the CO2 reforming of CH4 as a method of CO2 mitigation. A thermodynamic study. Prepr. Pap.-Am. Chem. Soc., Div. Fuel Chem., 49 (1), 126–127, (2004). CO2 Utilization: A Process Systems Engineering Vision http://dx.doi.org/10.5772/57560 87
92. Trusler, J. P. M. Equation of State for Solid Phase I of Carbon Dioxide Valid for Tem- peratures up to 800 K and Pressures up to 12 GPa. J. Phys. Chem. Ref. Data, 40 (4), 043105-1-043105-19, http://dx.doi.org/10.1063/1.3664915, (2011).
93. Van-Dal, E.S., Boualloub, C. Design and simulation of a methanol production plant from CO2 hydrogenation. Journal of Cleaner Production, 57, 15 October 2013, 38–45, http://dx.doi.org/10.1016/j.jclepro.2013.06.008
94. Ushikubo, T. Green Chemistry in Japan. Chemistry International, 35(4), July-August 2013. Available online at www.iupac.org/publications/ci, (2013).

95. Wagener, D.H.V., Rochelle, G.T. Alternative Stripper Configurations for CO2 Capture by Aqueous Amines. AIChE Spring Meeting, (2010).
96. Wagener, D.H.V., Rochelle, G.T. Stripper configurations for CO2 capture by aqueous monoethanolamine. Chemical Engineering Research and Design, 89, 1639–1646, (2011).
97. Walther, D., Ruben, M., Rau, S. Carbon Dioxide and Metal Centres: from Reactions Inspired by Nature to Reactions in Compressed Carbon Dioxide as Solvent. Coordi- nation Chemistry Reviews 182, 67–100, (1999).
98. Wang, W., Stagg-Williams, S.M., Noronha, F.B., Mattos, L.V.,Passos, F.B., Partial Oxi- dation and Combined Reforming of Methane on Ce-promoted Catalysts, Prepr. Pap.- Am. Chem. Soc., Div. Fuel Chem. 49 (1), 132-133, (2004).
99. Wu, J.C.S., Lin, H.-M., Photo reduction of CO2 to methanol via TiO2 Photocatalyst, In- ternational Journal of Photoenergy, 7, 115-119, (2005).
100. Wurzel, T., Malcus, S., Mleczko, L., Reaction engineering investigations of CO2 re- forming in a fluidized-bed reactor. Chem. Eng. Sci., 55, 3955-3966 (2000).
101. Yamashita, H., Fujii, Y., Ichihashi, Y., Zhang, S. G., Ikeue, K., Park, D. R.,Koyano, K., Tatsumi, T., Anpo, M., Selective formation of CH3OH in the photocatalytic reduction of CO2 with H2O on titanium oxides highly dispersed within zeolites and mesoporous molecular sieves, Catalysis Today 45, 221-227, (1998).
102. Zangouei, M., Moghaddam, A.Z., Razeghi, A., Omidkhah, M.R., Reforming and Par- tial Oxidation of CH4 over Ni/Al2O3 Catalysts in Fixed-bed Reactor International Journal of Chemical Reactor Engineering, 8. Available at: http://works.bepress.com/mohammadreza_omidkhah/2, (2010).
103. Zhang, X., Lee, C. S.-M., Mingos, D.M.P., Hayward, D. O., Carbon dioxide reforming of methane with Pt catalysts using microwave dielectric heating. Catalysis Letters, 88(3–4), 129-139, (2003).
104. Zhou, H., Cao, Y., Zhao, H., Liu, H., Wei-Ping, P., Investigation of H2O and CO2 Re- forming and Partial Oxidation of Methane : Catalytic Effects of Coal Char and Coal Ash, Energy & Fuels, 22, 2341-2345, (2008).

Chapter 6

A Future Perspective on the Role of Industrial Biotechnology for Chemicals Production

John M. Woodley[a], Michael Breuer[b], and Daniel Mink[c]

[a]Center for Process Engineering and Technology, Department of Chemical and Biochemical Engineering, Technical University of Denmark, DK-2800 Lyngby, Denmark
[b]BASF SE Fine Chemicals and Biocatalysis Research GVF/E - A030, D-67056 Ludwigshafen, Germany
[c]DSM Chemtech Center, 6160 MD Geleen, The Netherlands

ABSTRACT

The development of recombinant DNA technology, the need for renewable raw materials and a green, sustainable profile for future chemical processes have been major drivers in the implementation of industrial biotechnology. The use of industrial biotechnology for the production of chemicals is well established in the pharmaceutical industry but is moving down the value chain toward bulk chemicals. Chemical engineers will have an essential role in the development of new processes where the need is for new design methods for effective implementation, just as much as new technology. Most interesting is that the design of these processes relies on an integrated approach of biocatalyst and process engineering.

INTRODUCTION

This year the European Federation of Chemical Engineering (EFCE) celebrates sixty years of contributions to the chemical industry and the (technical) universities that support that industry. The last sixty years have seen enormous developments in the chemical industry and the discipline of chemical engineering, as reflected by the other papers in this special issue, and Europe has had an important role to play in these developments. By strange coincidence this year also marks the sixtieth anniversary of the discovery of the structure of deoxyribonucleic acid (DNA) (the double-helix) by Watson and Crick in Cambridge (UK). It is this discovery (for which they were jointly awarded the Nobel Prize in 1962, together with Maurice Wilkins) which, more than any other, has led to the development of a new branch of chemical engineering. Today, industrial biotechnology is a major global industry and chemical engineers have not only contributed to this astonishing development but will become increasingly involved in the future too. Interestingly bioprocessing is an area of innovation that leads not only to improved processes for existing products but also a range of entirely new products. Since the discovery of the structure of DNA, it took a further twenty years for molecular biologists to develop the set of tools we today refer to as recombinant DNA technology. Such tools enable access to new expression hosts, overexpression of genes, overproduction of proteins and pathway manipulation, as well as the design of proteins with new properties (Table 1). It is this development which has been such a huge driver for industrial biotechnology enabling new routes to chemical products via cheaper and tailor made biocatalysts (whether in growing cell, resting cell or isolated enzyme format). A second driver is the need to replace fossil resources and over time move to renewable raw materials. An increasing recognition that we have limited resources of fossil-based feed-stocks, metals and other reagents means that a new wave of processes is being developed based on sugar and vegetable oils as well as waste oils and fats. Such feed-stocks are very well suited to bio-based processing methods, since the molecules are already highly functionalized. The third driver is the need for green (clean) processes, where minimum waste is produced as well as efficient use is made of the energy used in producing the product. All three drivers (i.e. rDNA technology, raw material change and environmental impact) contribute ultimately to the overall economics of the process, and hence product. Obviously it will usually be the most economical solution to a given synthetic problem which will be implemented. Indeed a route with few steps and low environmental impact has the highest chance of success, no matter which technology is employed. In this brief article the motivation for the further development of industrial biotechnology for chemicals production will be discussed together with an outline of the specific role to be played by chemical engineers.

Table 1: Process implications of rDNA technology

Tool	Process benefits
Overexpressed protein	Cheaper biocatalyst production
Transfer between hosts organisms	Cheaper biocatalyst production
Alteration of cellular pathways	Designer biocatalysts (reactions)
Insertion of new pathways	Designer biocatalysts (reactions)
Protein engineering	Designer biocatalysts (properties)

TYPES OF BIOPROCESS

There are of course many ways of classifying the range of chemical and biochemical processes of potential use in manufacturing industry today to produce the enormous diversity of products available and useful for society. However one instructive classification is to analyze the nature of the feed-stock, reaction catalyst and product. Fig. 1 illustrates the range of conceptual conventional and new processes possible. Conventional processes use fossil-based feed-stocks and chemical catalysts to produce chemical products. Fig. 1 also shows that chemicals can be made with chemical catalysts but using new routes from renewable feed-stocks. This is an increasingly important area of technology, although beyond the scope of this article. However, with a biological catalyst (either using bio-catalysis or fermentation) the options are increased. For example the feed-stock can be fossil-based but may alternatively be renewable. Likewise the product may be a chemical or alternatively a larger biomolecule (an enzyme, protein, antibody or peptide). These so-called 'bio-products' (including biopharmaceutical, biologic and bio-similar products) have an important market but the scope of this paper will be solely on chemical products. Interestingly, the process to produce some products can be classified in more than one concept. For example biodiesel (the fatty acid alky ester resulting from the trans-esterification of vegetable oil) is considered a bio-based chemical. The raw material feed-stock for the manufacture of biodiesel can come from highly renewable resources (such as waste oil or algae), all the way to refined vegetable oils (in competition sometimes with food – renewable but not sustainable). Furthermore the reaction catalyst can be either enzymatic, chemical (alkaline) or a combination. Even the co-substrate alcohol (methanol or ethanol) could in the case of ethanol be from a renewable source (i.e. bio-ethanol via fermentation of sugar or lignocellulose) (Severson et al., 2013). This example illustrates well the complexity of classifying such systems. Nevertheless it is instructive to make some type of classification. In the future it will be increasingly important in order to understand the drivers for changing from conventional systems. Many of these drivers for change will set the agenda for the next generation of manufacturing processes. In this paper the scope is solely about concepts 3 and 4 (Fig. 1).

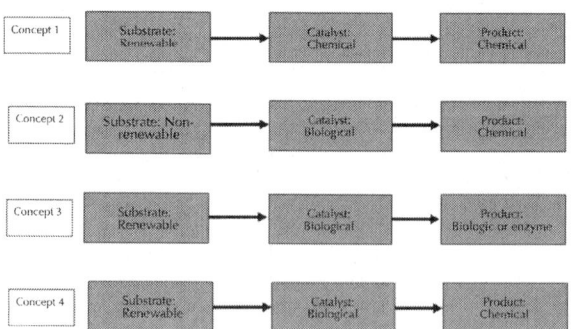

Figure 1: Potential classification of bioprocesses.

DRIVERS FOR CHANGE

Recombinant DNA Technology

The development of recombinant DNA (rDNA) technology enables several possibilities for the real exploitation of biocatalysts in the full sense of the word. First it has provided a cheap way to produce a given biocatalyst. The desired enzyme (or enzymes) can now be overproduced meaning that it represents a much bigger fraction of the available protein in the cell. This not only reduces the required scale of the fermentation (and consequently the feed-stock and energy required, as well as the waste produced) but for isolated enzyme applications also reduces the downstream burden prior to catalysis. Secondly a given gene may be expressed in vivo only in a poor host for production (e.g. the host may be pathogenic or grows only under conditions far from those used for application). Such a situation can be overcome by genetic engineering through codon optimization and subsequent cloning the desired gene into a new more appropriate host organism. Typical hosts in an industrial setting are for example *Bacillus subtilis*, *Escherichia coli*, *Aspergillus niger* or *Pichia pastoris* since they are fast growing (overcoming the risk of contamination) and the genetics are well understood, although others are used, dependent upon application. In some cases protein secretion is possible (e.g. from *Aspergillus niger*). Each of these developments has helped to revolutionize the biotechnology industry since they enable proteins and biocatalysts to be provided at a reasonable cost. The ability to grow cells to a high concentration (high cell 'density') based on sophisticated fed-batch feeding profiles has also had a major impact. Recombinant DNA technology can also enable an alteration of the properties of the biocatalyst. For cells this can involve alteration of pathways (blocking non-productive routes) and increasing metabolic flux or even creating *de-novo* pathways (Jones Prather and Martin, 2008 and Meyer et al., 2007). Today synthetic biology is a hugely exciting area of industrial biotechnology which will develop entirely new routes to chemicals. Whether it is to be carried out inside or outside the cell is still an open question. In some cases compartmentalization is useful and in other cases not. For enzymes, the ability to swap the amino acids either in the active site or even at remote positions of the protein has been found capable of altering and controlling substrate repertoire, enzyme stability, activity (reaction rate) and selectivity. Today, protein engineers routinely use so-called 'directed evolution' as well as rational strategies based on protein structure information to optimize proteins (Strohmeier et al., 2011 and Reetz, 2013). For the future this will be applied in processes at full scale. Examples already exist but it is clear this is a very exciting area that will develop enormously in the coming decades. The input of chemical and biochemical engineers is highly important here, because they alone can set the agenda for protein engineers, dependent upon process requirements.

Renewable Raw Materials and Reagents

A major preoccupation of the chemical industry in the last decade has been the development of alternative chemical production processes based on renewable feed-stocks, meaning those not derived from fossil or oil-based chemicals (Gwehenberger and Narodoslawsky, 2008). Such a shift is driven by increasing oil prices as well as the need for reducing dependence on a single source. Ultimately oil will run out and need replacing, and although there remains time till this takes place planning for the change now is important. The argument has mostly focused on the need for an alternative fuel source, which brings with it the huge challenge of making a product which can only be sold at a very low price. One approach has been the development of a so called 'bio-refinery' that supplements the cost of producing low value biofuel with higher value bio-based chemicals. In recent years the US Department of Energy has even defined a list of the top building block chemicals of importance for a bio-refinery based on renewable carbohydrates (Werpy and Petersen, 2004). For example succinic acid is one such chemical on the verge of commercialization. While it is not clear where such a debate will end it is evident that the use of carbohydrates as a feed-stock implicates fermentation and biocatalysis as likely, even if not the only, methods in producing such chemicals, complemented by chemo-catalysis (Bozell and Petersen, 2010, Dapsens et al., 2012, Thomas et al., 2002, Marr and Liu, 2011 and Herrera, 2004).

Green Processes

Good chemical engineering has long been recognized as that required to create, develop and design cost-effective processes for chemical manufacture. In the last sixty years the demands on manufacturing have developed such that today a second set of requirements is that processes have also to meet strict safety requirements, both for those involved in the manufacture of such products as well as those that use the products. It has been an important development, but now in the next sixty years a third requirement will be integrated with these – the environmental (and with it the societal) impact of the process. To some extent the integration of this third requirement is already underway but it needs further development. For example, aside from the driver toward renewable feed-stocks and reagents, there is also a trend toward manufacturing processes with minimal waste and which use reagents in a very effective way (Gwehenberger and Narodoslawsky, 2008). Ultimately this will lead to sustainable chemical processes and products. Industrial biotechnology is one tool available to achieve this although it is clearly not the only one. However a necessity in developing sustainable processes in the future will be that appropriate measurements are made of the 'eco-footprint' of the process. Today many metrics are used to assess such processes (Dreyer et al., 2003, Saling et al., 2002 and Andraos, 2013), from full life-cycle analysis (LCA) to green chemistry metrics (GCMs) (e.g. PMI; Kjell et al., 2013) or E-factor (kg waste/kg product) (Sheldon, 2007). While a range of metrics, models and software tools are available today (Shonnard et al., 2003 and Jiménez-González et al., 2013), they need to be developed further (and potentially standardized) to match every stage of development and be used to guide development and process improvement. For example the tools to be used at an early conceptual stage when limited data are available are clearly different to those to be used at a late stage of development, when detailed design is carried out. It is frequently argued that bioprocesses are cleaner than their chemical counterparts, but aside from the few isolated examples where comparisons have been undertaken (Vink et al., 2010 and Henderson et al., 2008), this argument must not be generalized. Many bioprocesses are highly selective and operate with low (Henderson et al., 2008) energy requirements (during operation) but are heavy on water usage and downstream processing. For the future making such comparisons will be important to build up a significant database of cases, such that expectations can be met (Jiménez-González et al., 2011 and Adlercreutz et al., 2010). Likewise integration into existing chemical factories may come at an environmental price due to the differing conditions required for chemical and biological reactions (Wenda et al., 2011).

CURRENT BIOPROCESS TECHNOLOGY

Although many of the principles of biochemical engineering are identical to those of chemical engineering, involving transport phenomena, thermodynamics, reaction rate laws and separation principles, combined with mass and energy balances, there are several features of bioprocesses which are special (Yuryev and Liese, 2010). It is these special features of industrial biotechnology which perhaps more than anything else distinguish biochemical engineering as a separate discipline.

- Unlike many chemical reactions (which often take place in the gas phase) many of the reactions which constitute the synthetic stage of bioprocesses take place in liquid phase, frequently water. This means that transport phenomena are different and it also has implications for the recovery of products from such media. This is particularly critical given that in nature bioprocesses work in dilute conditions, meaning that integration into existing processes is frequently difficult (or requires large investments). In all cases for industrial application it is necessary to engineer either the biocatalyst to cope with higher concentrations or the process to overcome low concentrations, or a combination of both. Given the need for high product concentration to keep the downstream and waste costs manageable, it will also be necessary to address higher substrate concentrations as well.
- Any heat produced in an exothermic reaction (e.g. in an aerobic fermentation) will be at ambient temperature (or maybe a little above). This means heat integration with the downstream process is difficult, since this is not useful heat.

- Frequently (and especially in fermentation) the product is not the major component to enter the downstream process. This situation places particular emphasis on the selection of effective unit operations in the downstream process.
- The interaction between unit operations is particularly strong, meaning that changes upstream (in the fermentation or biocatalysis stage) can have implications further downstream, where the product is recovered, purified and polished. Indeed often there are changes downstream simply because the output from the fermentation is not consistent.

Table 2 lists these key features and the implications for engineering the process and the biocatalyst. It is clear that chemical and biochemical engineers have an important role to play in addressing these process problems. This also places particular emphasis on the development processes itself, where chemical engineers need to work hand-in-hand with biochemists and microbiologists at an early stage of development.

Table 2: Process features of biocatalytic processes

Advantages
Selective chemistry
Mild condition reaction conditions
Water based reactions, minimizing organic solvent use
Integration with other bioprocessing steps due to compatible reaction conditions
Disadvantages
Catalyst and process needs to be engineered to match industrial requirements
Catalyst need to be engineered to enable conversion of non-natural substrates
Liquid based reactions
Integration with chemical steps due to incompatability of reaction conditions

Industrial Fermentation Processes

Industrial fermentation is a multi-step processes where cultivation is increased in scale in a staged approach. Today most industrial fermentation products are produced at the largest scale via fed-batch strategies. Limiting a carbon source in this way by feeding is used to slow down the fermentation in the latter stages, with the advantage that oxygen supply and heat removal can be managed. Nevertheless for aerobic fermentations the size of an individual unit is limited to around 200 m^3, making it hard to gain the benefits of economies of scale seen in conventional petrochemical processes. For anaerobic fermentation the yield of product is often, although not always, lower (since more of the carbon goes into the cells and other products). Nevertheless, final the scale can be as high as 500–1000 m^3. A plethora of chemical products (Miller and Nagarajan, 2000, Whited et al., 2010 and Weusthuis et al., 2011) are produced by fermentation today including organic acids and alcohols. Limitations in the further development of such processes are the cost and availability of feed-stocks as well as the genetic engineering of the cells (Straathof et al., 2005). Ultimately energy input, product yield on substrate, space-time-yield and product concentration will determine the economics of such processes.

Industrial Biocatalytic Processes

Bioprocesses based on the use of one or more enzymes as a catalyst, which may be in non-growing cells (often referred to as 'resting' cells) or supported on an 'immobilization' matrix or soluble in solution are known as biocatalytic processes (Schoemaker et al., 2003). Such processes are most prevalent today in the pharmaceutical industry (Pollard and Woodley, 2007) but also have a major impact in many other areas of chemical production, such as oils and fats processing (Schmid et al., 2001). Around 200 processes alone are already implemented in the pharmaceutical industry. Some of the issues surrounding industrial application of biocatalytic processes are summarized in Table 3. Limitations in the further development of such processes are the cost of downstream processing as well as the protein engineering of the enzymes. Ultimately product yield on substrate, product yield on enzyme, space-time-yield and product concentration determine the economics of such processes.

Table 3: Examples of potential challenges in the industrial application of biocatalytic processes

Availability of some biocatalysts at industrially relevant scale.
Selection of biocatalyst format (e.g. isolated enzyme, immobilized enzyme, whole-cell).
Time taken to 'evolve' or screen the biocatalyst to operate at the required rate with non-natural substrates at the appropriate substrate and product concentrations.
Time taken to integrate process engineering solutions with the biocatalyst properties to achieve 'optimal' process configuration and operating strategy.

Downstream Processing

The recovery of chemical products from fermentation and microbial-based or enzyme-based biocatalysis is often expensive and difficult. The primary reason is the dilute nature of the product stream. For a commercial process, concentrations around 100 g/L are required for high value products and around 300–400 g/L for bulk chemicals. Such concentrations are far away from those found in nature and although some biocatalyst engineering solutions may help, the low concentration is frequently the result of the inhibitory or toxic effects of products or co-products (or the dilute nature of the catalyst preparation). A potential solution developed over the last thirty years is *in situ* product removal (ISPR) (or *in situ* by-product removal – ISBR), where the product (or by-product) is removed from the site of reaction as is it formed (seeFig. 2). A variety of techniques and technologies are available today (Freeman et al., 1993, Stark and von Stockar, 2003 and Woodley et al., 2008). It seems likely that such technology will be further developed in the future. Process integration of this type, where unit operations are combined has many complexities for biotechnological processes, but if mastered will enable many lower value products to be manufactured using biotechnological means. One of the most important new technologies for industrial biotechnology involves the membrane based unit-operations. Such operations are required especially for the compartmentalization and/or recycle of enzyme or cells. In many cases contactors to enable liquid-liquid extraction are also possible, also via in situ product removal (Pabby and Sastre, 2013).

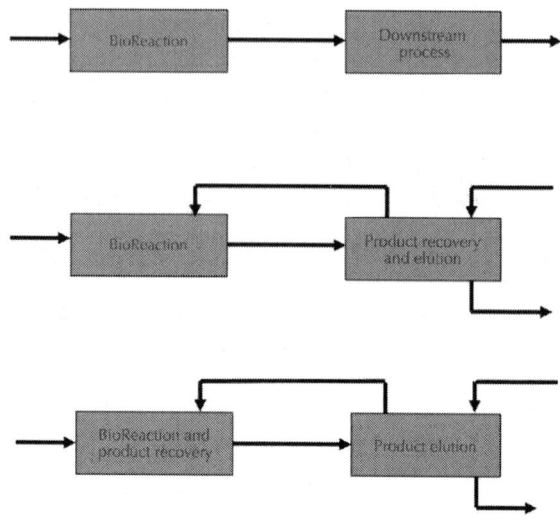

Figure 2: Schematic ISPR options.

FUTURE BIOPROCESSES

Process Integration with Chemical Operations

It is clear that for the future it will be necessary to integrate bioprocesses into existing chemical plants. Today the cost of many chemical plants is already written-off, so replacement is not an easy option. This is particularly true for large plants of at least 500 million investment. In stages it seems likely that (1) capacity increases will be absorbed by bioprocesses and later (2) retrofit of existing processes will be carried out. At first a significant majority of steps will remain chemically based and later bioprocesses will expand into the majority if not the entire process. Finally it is clear that bioprocesses which operate under mild conditions (neutral pH, atmospheric pressure and ambient temperature) will enable options for process plant made from cheaper materials of construction. In some areas of bioprocessing disposable plant and polypropylene based tanks and pipes are already being used. Clearly this will have important consequences for plant design and investment of capital.

Next Generation Biocatalysts

Biocatalysts of the future will be far more sophisticated than today. For example rather than using a single enzyme for a conversion, multiple enzymes will be used in entirely new pathways or routes (Santacoloma et al., 2011, Sagt, 2013, Gardner, 2013 and Xue and Woodley, 2012). Already today such an approach is in use for in situ cofactor regeneration. Such a concept builds on what nature already provides today (Bruggink et al., 2003). While this can provide inspiration, new pathways are required to build the necessary products of the future and attention will need to be paid not only to kinetics but also the thermodynamics of such de novo pathways. On occasion there will be value in operating the pathways outside cells and on other occasions inside cells (Rollié et al., 2012). In both cases the yield will be required to be focused on the desired product, with the aim of reducing intermediate separation and losses to by-products. Likewise protein engineering (Lutz, 2010 and Bornscheuer et al., 2012) will enable alteration to biocatalysts such that designer catalysts will become possible (potentially enabling new chemistries), ultimately ensuring integrated operation in one pot, eliminating entirely intermediate separation and in some cases linked to neighboring chemical reaction steps (Woodley, 2013 and Wei, 1996). Finally, there are already a few precedents where enzymes have not just been modified, but also designed from scratch capable of catalyzing reaction types which have not been found in vivo (Röthlisberger et al., 2008 and Siegel et al., 2010)

FUTURE ENGINEERING TOOLS

Rapid Design Methods

Biology does not provide all the necessary solutions for industrial process challenges. The last decades have clearly shown the need for innovative process engineering solutions as well as highly sophisticated molecular biology to engineer the biocatalyst. Indeed it is highly likely integration of catalyst design and process design is part of the next paradigm in chemical engineering (Vennestrøm et al., 2010). Particularly interesting is that neither a single objective (e.g.

lowest production cost or lowest development cost), nor development route (e.g. protein engineering or process engineering) nor solution (e.g. enzyme immobilization for 100 recycles with microfiltration or 5 recycles with soluble enzyme by ultrafiltration) exist, even in a given case. An excellent recent industrial example from the pharmaceutical industry is the use of an engineered -transaminase (EC 2.6.1.18) for the manufacture of Sitagliptin by Merck (USA), which was solved by integrated process and biocatalyst engineering (Savile et al., 2010 and Truppo et al., 2012). In other sectors of industrial biotechnology the economic leeway for process optimization might be more limited than in the pharmaceutical business, but there remains room for improvement in many cases.

Frequently multiple strategies are required to come to a satisfactory economic solution. A clear need from the perspective of process engineers is to develop a means to navigate the solution space in an effective way. In the pharmaceutical sector the time limitations (as a result of a defined patent life) mean that the emphasis is on speed of development. It is clear that automated, systematic methods of data collection, linked with design of experiments and process models will have huge benefits in much the way they have already for in other sectors of the chemical industry.

Process Modeling

Process modeling is increasingly implemented as a means of mathematically describing bioprocesses. It is of course easiest for enzyme-based bio-catalysis (Vasi -Ra ki et al., 2011), but is also necessary for complex fermentations where population based models are required. Two types of models need to be developed – those that describe the reaction phase and those that describe the downstream unit operations. Much progress has already been made but more sophisticated models are required to enable a more predictive approach for scale-up and design. This will also be an important contribution from chemical engineers in the future as we move from empirical to more mechanistically based models. Alongside this it will be necessary to build property databases of suitable feed-stocks, reagents, and chemicals. In many cases predictive tools for the properties of many of these molecules would perhaps be even more useful, to save valuable experimental time. The chemistry, in particular of many processes where biological catalysts can best be exploited, is complex and the building of a suitable database and predictive tools will be an important contribution.

FUTURE PERSPECTIVES

There is little doubt that industrial biotechnology will expand as an important area of chemical engineering for the future. However this also represents a challenge both for academia and also industry, where change from the *status quo* will be required.

- *Changing academia.* In academia the challenge is how to educate chemical engineers in industrial biotechnology. Already today the curriculum is filled so full covering the basic scientific background and the special principles of chemical engineering as well as design. Many degree programs also offer extra courses that enable some degree of specialization from polymer technology to pharmaceutical processing. So one can easily conclude that there seems little room in the existing curriculum for teaching about bioprocesses and it is clear that master's programs focusing on this area are required. Many universities already offer such programs although here too the curriculum can become filled with the enormous diversity of bioprocesses from protein processing to small molecule processing. So even here some specialization may be required. The support of industry in enabling exposure to real industrial problems (via work experience or even case studies) is also important, but will require significant commitment. A final issue is that the very rapid development of the discipline has led also to a shortage of definitive texts for use as a support for students. Making biology quantitative will also require special care both from those teaching and those being taught.
- *Changing industry.* Much of the chemical industry today is under pressure. The need to innovate is one solution but it is still not straightforward. For example, in order to innovate via introduction of bioprocesses, various companies have tried different approaches ranging from specialist in-house teams to outsourcing, but in the end integration into the mainstream of a company from managers to design engineers means that new approaches are required including much more active continuing

education programs to help and assist industry. To ensure success in the future, close collaboration, which is already a hallmark of industrial biotechnology today, will be essential between chemical engineers, chemists and biologists, preferably in a single physical location.

CONCLUDING REMARKS

The last sixty years have seen enormous progress in the development of industrial biotechnology for chemical production and today many companies using such processes in an effective way to produce chemicals. Indeed today we can state that it is no longer a new technology but one that is taking a mainstream role in many research and development organizations in industry. Nevertheless education of the next generation of engineers and the continuing education of those already working in industry is also required. In such developments Europe has a particularly strong position and partly this is as a result of supportive funding for research from the EC and partly the result of excellent collaboration between industry and academia. This not only represents an exciting development but also helps all those in university to understand industrial needs and for industry to pick the very latest developments from academia. Such a synergistic relationship has served bioprocessing very well in the past six decades and there is little doubt it will continue to do so in the next six as well.

REFERENCES

1. Adlercreutz, D., Tufvesson, P., Karlsson, A., Hatti-Kaul, R., 2010. Alkonolamide biosurfactants: techno-economic evaluation of biocatalytic versus chemical production. Ind. Biotechnol. 6, 204–211.
2. Andraos, J., 2013. Safety/hazard indices: completion of a unified suite of metrics for the assessment of 'Greeness' for chemical reactions and synthesis plans. Org. Proc. Res. Dev. 17, 175–192.
3. Bornscheuer, U.T., Huisman, G.W., Kazlauskas, R.J., Lutz, S., Moore, J.C., Robins, K., 2012. Engineering the third wave of biocatalysis. Nature 485, 185–194.

4. Bozell, J.J., Petersen, G.R., 2010. Technology development for the production of biobased products from biorefinery carbohydrates—the US Department of Energy's Top 10 revisited. Green Chem. 12, 539–554.
5. Bruggink, A., Schoevaart, R., Kieboom, T., 2003. Concepts of nature in organic synthesis: cascade catalysis and multistep conversions in concert. Org. Proc. Res. Dev. 7, 622–640.
6. Dapsens, P.Y., Mondelli, C., Pérez-Ramírez, J., 2012. Biobased chemicals form conception toward industrial reality: lessons learned and to be learned. ACS Catal. 2, 1487–1499.
7. Dreyer, L.C., Niemann, A.L., Hauschild, M.Z., 2003. Comparison of three different LCIA methods: EDIP97, CML2001 and Eco-indicator 99. Int. J. LCA 8, 191–200.
8. Freeman, A., Woodley, J.M., Lilly, M.D., 1993. In situ product removal as a tool for bioprocessing. Biol. Technol. 11, 1007–1012.
9. Gardner, T.S., 2013. Synthetic biology: from hype to impact. Trends Biotechnol. 31, 123–125.
10. Gwehenberger, G., Narodoslawsky, M., 2008. Sustainable processes–the challenge of the 21st century for chemical engineering. Process Safety Environ. Protect. 86, 321–327.
11. Henderson, R.K., Jimenez-Gonzalez, C., Preston, C., Constable, D.J.C., Woodley, J.M., 2008. Comparison of biocatalytic and chemical synthesis: EHS and LCA comparison for 7-ACA synthesis. Ind. Biotechnol. 4, 180–192.
12. Herrera, S., 2004. Industrial biotechnology—a chance at redemption. Nat. Biotechnol. 22, 671–675.
13. Jiménez-González, C., Poechlauer, P., Broxterman, Q.B., Yang, B S, am Ende, D., Baird, J., Bertsch, C., Hannah, R.E., Dell'Orco, P., Noorman, H., Yee, S., Reintjens, R., Wells, A., Massonneau, V., Manley, J., 2011. Key green manufacturing research areas for sustainable manufacturing: a perspective form pharmaceutical and fine chemicals manufacturers. Org. Proc. Res. Dev. 15, 900–911.
14. Jiménez-González, C., Ollech, C., Pyrz, W., Hughes, D., Broxterman, Q.B., Bhathela, N., 2013. Expanding the boundaries: developing a streamlined tool for eco-footprinting of pharmaceuticals. Org. Proc. Res. Dev. 17, 239–246.

15. Jones Prather, K.L., Martin, C.H., 2008. De novo biosynthetic pathways: rational design of microbial chemical factories. Curr. Opin. Biotechnol. 19, 468–474.
16. Kjell, D.P., Watson, I.A., Wolfe, C.N., Spitler, J.T., 2013. Complexity-based metric for process mass intensity in the pharmaceutical industry. Org. Proc. Res. Dev. 17, 169–174.
17. Lutz, S., 2010. Beyond directed evolution—semi-rational protein engineering and design. Curr. Opin. Biotechnol. 21, 734–743.
18. Marr, A.C., Liu, S., 2011. Combining bio- and chemo-catalysis: from enzymes to cells, from petroleum to biomass. Trends Biotechnol. 29, 199–204.
19. Meyer, A., Pallaux, R., Panke, S., 2007. Bioengineering novel in vitro metabolic pathways using synthetic biology. Curr. Opin. Microbiol. 10, 246–253.
20. Miller, J.A., Nagarajan, V., 2000. The impact of biotechnology on the chemical industry in the 21st century. Trends Biotechnol. 18, 190–191.
21. Pabby, A.K., Sastre, A.M., 2013. State-of-the-art review on hollow fibre contactor technology and membrane-based extraction processes. J. Membr. Sci. 430, 263–303.
22. Pollard, D.J., Woodley, J.M., 2007. Biocatalysis for pharmaceutical intermediates: the future is now. Trends Biotechnol. 25, 66–73.
23. Reetz, M.T., 2013. The importance of additive and no-additive mutational effects in protein engineering. Angew. Chem. Int. Ed. Engl. 52, 2658–2666.
24. Rollié, S., Mangold, M., Sundmacher, K., 2012. Designing biological systems: systems engineering meets synthetic biology. Chem. Eng. Sci. 69, 1–29.
25. Röthlisberger, D., Khersonsky, O., Wollacott, A.M., Jiang, L., De Chancie, J., Betker, J., Gallaher, J.L., Althoff, E.A., Zanghellini, A., Dym, O., Albeck, S., Houk, K.N., Tawfik, D.S., Baker, D., 2008. Kemp elimination catalysts by computational enzyme design. Nature 453, 190–195.
26. Sagt, C.M.J., 2013. Systems metabolic engineering in an industrial setting. Appl. Microbiol. Biotechnol. 97, 2319–2326.
27. Saling, P., Kicherer, A., Dittrich-Krämer, B., Wittlinger, R., Zombik, W., Schmidt, I., Schrott, W., Schmidt, S., 2002. Eco-efficiency analysis by BASF: the method. Int. J. LCA. 7, 203–218.

28. Santacoloma, P.A., Sin, G., Gernaey, K.V., Woodley, J.M., 2011. Multi-enzyme catalyzed processes: next generation biocatalysis. Org. Proc. Res. Dev. 15, 203–212.
29. Savile, C.K., Janey, J.M., Mundorff, E.C., Moore, J.C., Tam, S., Jarvis, W.R., Colbeck, J.C., Krebber, A., Fleitz, F.J., Brands, J., Devine, P.N., Huisman, G.W., Hughes, G.J., 2010. Biocatalytic asymmetric synthesis of chiral amines from ketones applied to sitagliptin manufacture. Science 329, 305–309.
30. Schmid, A., Dordick, J.S., Hauer, B., Kiener, A., Wubbolts, M., Witholt, B., 2001. Industrial biocatalysis today and tomorrow. Nature 409, 258–268.
31. Schoemaker, H.E., Mink, D., Wubbolts, M.G., 2003. Dispelling the myths—biocatalysis in industrial synthesis. Science 299, 1694–1697.
32. Severson, K., Martín, M., Grossmann, I.E., 2013. Optimal integration for biodiesel production using bioethanol. AIChE J. 59, 834–844.
33. Sheldon, R.A., 2007. The E-factor fifteen years on. Green Chem. 9, 1273–1283.
34. Shonnard, D.R., Kircherer, A., Saling, P., 2003. Industrial applications using BASF eco-efficiency analysis: perspectives on green engineering principles. Environ. Sci. Technol. 37, 5340–5348.
35. Siegel, J.B., Zanghellini, A., Lovick, H.M., Kiss, G., Lambert, A.R., St Clair, J.L., Gallaher, J.L., Hilvert, D., Gelb, M.H., Stoddard, B.L., Houk, K.N., Michael, F.E., Baker, D., 2010. Computational design of an enzyme catalyst for a stereoselective bimolecular Diels–Alder reaction. Science 329, 309–313.
36. Stark, D., von Stockar, U., 2003. In-situ product removal (ISPR) in whole cell biotechnology during the last 20 years. Adv. Biochem. Eng. Biotechnol. 80, 149–175.
37. Straathof, A.J.J., Sie, S., Franco, T.T., van der Wielen, L.A.M., 2005. Feasibility of acrylic acid production by fermentation. Appl. Microbiol. Biotechnol. 67, 727–734.
38. Strohmeier, G.A., Pichler, H., May, O., Gruber-Khadjawi, M., 2011. Application of designed enzymes in organic synthesis. Chem. Reviews 111, 4141–4164.

39. Thomas, S.M., DiCosimo, R., Nagarajan, V., 2002. Biocatalysis: applications and potentials for the chemical industry. Trends Biotechnol. 20, 238–242.
40. Truppo, M.D., Strotman, H., Hughes, G., 2012. Development of an immobilized transaminase capable of operating in organic solvent. ChemCatChem 4, 1071–1074.
41. Vasic-Ra´cki, ˇD., Findrik, Z., Vrsalovic´ Presecki, ˇA., 2011. Modeling as a tool of enzyme reaction engineering for enzyme reactor development. Appl. Microbiol. Biotechnol. 91, 845–856.
42. Vennestrøm, P.N.R., Christensen, C.H., Pedersen, S., Grunwaldt, J-D., Woodley, J.M., 2010. Next generation catalysis for renewables: combining enzymatic with inorganic heterogeneous catalysis for bulk chemical production. ChemCatChem 2, 249–258.
43. Vink, E.T.H., Davies, S., Kolstad, J.J., 2010. The eco-profile for current Ingeo® polylactide production. Ind. Biotechnol. 6, 212–224.
44. Wei, J., 1996. A century of changing paradigms in chemical engineering. Chem. Technol. 26, 16–18.
45. Wenda, S., Illner, S., Mell, A., Kragl, U., 2011. Industrial biotechnology—the future of green chemistry? Green Chem. 13, 3007–3047.
46. Werpy, T., Petersen, G., 2004. Top Value Added Chemicals from Biomass. US Department of Energy, Office of Scientific and Technical Information. DOE/GO-102004-1992 www.osti.gov/bridge
47. Weusthuis, R.A., Lamot, I., van der Oost, J., Sanders, J.P.M., 2011. Microbial production of bulk chemicals: development of anaerobic processes. Trends Biotechnol. 29, 153–158.
48. Whited, G.M., Fehrer, F.J., Benko, D.A., Cervin, M.A., Chotani, G.K., MicAuliffe, J.C., LaDica, R.J., Ben-Shoshan, E.A., Sanford, K.J., 2010. Development of a gas-phase bioprocess for isoprene-monomer production using metabolic pathway engineering. Ind. Biotechnol. 6, 152–163.
49. Woodley, J.M., 2013. Protein engineering of enzymes for process applications. Curr. Opin. Chem. Biol. 17, 310–316.

50. Woodley, J.M., Bisschops, M., Straathof, A.J.J., Ottens, M., 2008. Future directions for in-situ product removal (ISPR). J. Chem. Technol. Biotechnol. 83, 121–123.
51. Xue, R., Woodley, J.M., 2012. Process technology for multi-enzymatic reaction systems. Bioresour. Technol. 115, 183–195.
52. Yuryev, R., Liese, A., 2010. Biocatalysis: the outcast. ChemCatChem 2, 103–107.

Chapter 7

Safety Analysis Approach Based on Thermodynamic and Chemical Reactions Modelling

Taha Benikhlef[a], Djamel Benazzouz[a], Smail Adjerid[a], and Kazimierz Lebecki[b]

[a]Laboratoire de Mécanique des Solides et Systèmes (LMSS), Université M'Hamed Bougara, Boulevard de l'indépendance, Boumerdès, Algeria
[b]School of Occupational Safety Management, Katowice, Poland

ABSTRACT

Safety analysis of nuclear and chemical/petrochemical facilities is the systematic process that is carried out throughout the design process to ensure that all the relevant safety requirements are met by the proposed design of the plant. Safety analysis should incorporate both deterministic and probabilistic approaches. These approaches have been shown to complement each other and both should be used in the decision making process on the safety and ability of the plant to be licensed.

This paper deals with the deterministic safety approach in order to distill the experience of nuclear and chemical/petrochemical protection engineering through a safety analysis approach aiming at analysis of chemically reacting processes including thermodynamic and chemical reactions modelling that are present in both industries. For instance, there are some similarities between the Bhopal disaster and Three Mile Island-Fukushima-like H_2 deflagration-detonation scenarios in nuclear containments. The phenomenology is similar in that the temperature and the pressure caused by exothermic reactions had increased dramatically leading to a loss of containment.

The study aims to translate and adapt to general chemically reacting modelling, major features of the equivalent analysis inside the nuclear containments. Compartment containment for H_2 deflagrations has been translated and adapted, with fixed tools, to the methyl-isocyanate storage tank 610 of the Bhopal plant.

INTRODUCTION

Since the Bhopal disaster in 1984, a lot of attention has been paid to the chemical and petrochemical protection engineering field. Also a great deal of work related to protection engineering has been performed in the nuclear facilities before and after the TMI and Chernobyl accidents in 1979 and 1986.

Safety analysis of nuclear and chemical/petrochemical facilities is becoming a key to engineering activities, which is necessary to optimize protective layers to limit societal and individual risk. It usually involves two parts (Kirchsteiger, 1999):

- deterministic analysis based on dynamic process modelling with the goal of verifying the protection design (Medina, Arnaldos, & Casal, 2009).
- probabilistic analysis, aiming to assess the frequency of consequences of accidents outside the design basis (Labeau, Smidts, & Swaminathan, 2000).

Safety analysis will continue to grow in importance for many nations, not just in the European Union countries. As countries in North Africa and Middle East continue to develop their chemical, metallurgical, and pharmaceutical industries, health and safety analyses will take on greater importance to these nations.

Essential parts of deterministic safety are performed by applying sophisticated computer code packages which have been specifically developed for this purpose (Breitung & Royl, 2000). Usually, these code packages have been introduced into accident analysis of the nuclear industry through thermal-hydraulic tools analysis. As a relevant example, these tools have been developed and applied to loss of coolant accidents (Tregoures, Philippot, Foucher, Guillard, & Fleurot, 2010). They have also been broadly applied to assess safety of nuclear reactors with good prediction capabilities.

A parallel situation developed in the chemical and the petrochemical engineering, with less detailed codes in its thermal-hydraulic features, but incorporating the complexities of chemical reactions.

There are some similarities between Bhopal disaster, and TMI-Fukushima-like H_2 deflagration scenarios in nuclear containments. The phenomenology is similar in that the temperature and the pressure caused by exothermic reactions increased leading to a risk of loss of containment.

This study has been conducted to establish a common safety analysis platform between the safety analysis developed in the nuclear energy scientific community and the safety analysis which may affect the chemical and petrochemical plant processes.

The study shows the feasibility of a common tool analysis for scoping compartment containment analysis of H_2 deflagrations, able to translate and adapt the H_2 scenario to the methyl-isocyanate storage tank 610 explosion of Bhopal plant. No attempt is made to actually reproduce the Bhopal scenarios, rather to show how the tool may handle events of its kind.

DESCRIPTION OF THE CHEMICAL REACTIONS

Exothermic Reactions inside Tank 610

The Bhopal Gas Leak, India 1984 was the largest chemical industrial accident ever (Eckerman, 2005). About 520,000 persons were exposed to the gases, and it was estimated that 8000 died during the first weeks after the accident. About 100,000 persons or more received permanent injuries (Lepkowski, 1985). In the early hours of December 3, 1984, an estimated amount of 41 tons of deadly methyl-isocyanate (MIC) gas leaked out of tank 610 of the Union Carbide plant and escaped into the atmosphere. The immediate cause was the building up of pressure in the tank, due to an exothermic reaction caused by introduction of water inside the tank. This pressure caused the safety valve to rupture and the gas to escape.

Methyl-isocyanate (MIC) is reactive, toxic, volatile, and flammable. Methyl-isocyanate reacts exothermically with water to form 1,3-dimethylurea (a) and 1,3,5-trimethylbiuret (b) with evolution certain aspects of carbon dioxide (Themistocles, Anibal, Russell, Sureerat, & Chan, 1986). Excessive water leads to (a) and limited amounts of water (excess MIC) to (b).

- Excess water

$$\underset{MIC}{CH_3N = C = O} + H_2O \rightarrow CH_3NH_2 + CO_2$$

$$CH_3NH_2 + CH_3N = C = O \rightarrow \underset{(a)}{CH_3NHCONHCH_3}$$

- Excess MIC

$$\underset{(a)}{CH_3NHCONHCH_3} + CH_3N = C = O \rightarrow \underset{(b)}{CH_3NHCONCH_3CONHCH_3}$$

reacts further with methyl-isocyanate or other reaction products to give either 1,3-dimethylurea (with excess water) or 1,3,5-trimethylbiuret (with excess MIC). Another hypothesis, perhaps more plausible, is that methyl-isocyanate, in addition to reacting with a great many other substances, can also react with itself. In the presence of a catalyst, purified methyl-isocyanate will form either a cyclic trimer (trimethyl isocyanurate) or a gummy, resinous polymer.

The methyl-isocyanate water reaction in tank 610 starts off slowly with production of heat. If the heat is not somehow removed, the temperature will go up and the reaction rate will rapidly increase to the point that the methyl-isocyanate will start to boil violently. In a closed tank, the pressure could build up to the point that relief valves would open, venting both methyl-isocyanate vapour and carbon dioxide. If safety devices failed to operate, or if they were overwhelmed by the amount of vapours being generated, the heavy, noxious methyl-isocyanate vapours would escape to the atmosphere.

Combustion inside the Nuclear Reactor Containment

Hydrogen combustion in the nuclear containment building could produce pressure and temperature levels that threaten the integrity of the containment boundary.

The steam–zirconium reaction is the major source of hydrogen. At high temperature, metallic zirconium is then oxidized by the steam water to form hydrogen gas according to the following reaction (Berger, El Tahhann, Moulin, & Viennot, 2003):

$$Zr + 2H_2O \rightarrow ZrO_2 + 2H_2$$

During the accidental transient, hydrogen is released in the containment atmosphere. Hence, if the hydrogen concentration inside the containment is high enough, it reacts exothermically according to the following reaction:

$$H_2 + \frac{1}{2}O_2 \rightarrow H_2O$$

This reaction was responsible for a small hydrogen explosion accident first observed inside the reactor building of Three Mile Island nuclear power plant in 1979 (Peeters, 2007). This same reaction occurred probably in the boiling water reactors 1, 2 and 3 of the Fukushima Dai-Ichi nuclear power plant (Japan, 2011).

THERMODYNAMIC AND CHEMICAL REACTIONS MODELLING

Thermodynamic Properties Modelling

To determine the complete thermodynamic properties requires the sum of the ideal gas properties contribution and the residual correction for non-ideal behaviour (Vidal, 1997). Moreover, all the thermodynamic properties can be derived from the knowledge of equations of state, which interrelate the state variables of the system.

The use of cubic equations of state has become widespread because of their favourable mathematical characteristics, their simplicity, and their low computer time requirements (Hasan, Stanley, & Arvind, 1998). Many cubic equations of state like van der Waals, Redlich–Kwong, Redlich–Kwong–Soave, and Peng–Robinson are frequently used for phase equilibria calculations.

The following generalized cubic equation of state used to describe the behaviour of pressure can be written as:

$$p = \frac{RT}{V-b} - \frac{a_c m(T)}{V^2 + c_1 V + c_0} \quad (1)$$

where a_c value of the attraction parameter at the critical temperature.

Pressure Calculation

In this work, the Peng–Robinson equation of state has been applied to a pure fluids for both liquid and vapour phases. It is given by:

$$p = \frac{RT}{V-b} - \frac{a(T)}{V(V+b) + b(V-b)} \qquad (2)$$

With:

$$a(T) = a_c m(T)$$

where:

$$m(T) = \left(1 + \kappa\left(1 - \sqrt{\frac{T}{T_c}}\right)\right)^2;$$

$$a_c = 0.4572355 \frac{R^2 T_c^2}{P_c}; \quad b = 0.07779674 \frac{RT_c}{P_c}.$$

The parameter κ is related to Pitzer's acentric factor ω:

$$\kappa = 0.37464 + 1.5226\omega + 0.26992\omega^2$$

Calculation of Residual Properties

A residual quantity is the difference between a property of a real fluid and the same property of an ideal gas at the same density and temperature. The residual Helmholtz free energy Ar(T,V) is obtained as follows:

$$A^r(T,V) = -\int^V \left(p - \frac{RT}{V}\right)dV \qquad (3)$$

By substituting pressure p by Eq. (2), we write Eq. (3), in terms of dimensionless densities $\xi = b/V$ and dimensionless temperature $\bar{T} = T/T_s$:

$$\frac{A^r}{RT} = -\ln(1-\xi) - 2\sqrt{2}\frac{m(\bar{T})}{(\bar{T})}\ln\frac{1+\xi(\sqrt{2}+1)}{1-\xi(\sqrt{2}-1)} \qquad (4)$$

where T_s represents the characteristic temperature.

For convenience the Peng–Robinson equation is often written in a cubic polynomial form:

$$(8 - T' - p')\xi^3 + (2T' + 3p' - 8)\xi^2 + (T' - p')\xi - p' = 0 \qquad (5)$$

With:

$$p' = \frac{pb}{RT_s m(\bar{T})} \quad \text{and} \quad T' = \frac{\bar{T}}{m(\bar{T})}.$$

From the residual Helmholtz free energy, all other residual thermodynamic properties needed for thermodynamic functions can be derived. Therefore the residual Gibbs free energy can be written as:

$$\frac{G^r}{RT} = \frac{A^r}{RT} + Z - 1 \qquad (6)$$

Also, the residual internal energy can be written as:

$$\frac{U^r}{RT} = -T\left(\frac{\partial(A^r/RT)}{\partial T}\right)_V \qquad (7)$$

The residual enthalpy can then be written as:

$$\frac{H^r}{R} = \frac{U^r}{RT} + Z - 1 \qquad (8)$$

The residual isochoric and isobaric heat capacities can be written as:

$$\frac{C_V^r}{R} = -2\left(\frac{\partial(A^r/RT)}{\partial T}\right) - T\left(\frac{\partial^2(A^r/RT)}{\partial T^2}\right); \qquad (9)$$

And:

$$\frac{C_P^r}{R} = \frac{C_V^r}{R}\frac{\alpha_P^2}{\kappa_T} - 1 \qquad (10)$$

where α_P and κ_T represent respectively, the isobaric thermal expansion coefficient and the isothermal compressibility coefficient, being:

$$\alpha_p = \frac{1}{V}\left(\frac{\partial V}{\partial T}\right)_P \quad \text{and} \quad \kappa_T = \left(\frac{\partial p}{\partial \rho}\right)_T \tag{11}$$

Calculation of Ideal Gas Properties

The ideal isobaric heat capacity is obtained by using a polynomial form of the Shomate equation (NIST, 2008):

$$C_p^{id} = A + BT + CT^2 + DT^3 + E/T^2 \tag{12}$$

The constants A, B, C, D and E are given by NIST data bank. The ideal isochoric heat capacity can be written as:

$$C_V^{id} = C_p^{id} - R \tag{13}$$

To calculate the ideal enthalpy, we use the standard equation. For instance:

$$H^{id} = \int_{T_{ref}}^{T} C_p^{id} \, dT; \tag{14}$$

T_{ref} represents the reference temperature under standard conditions (T_{ref}=298.15 and P_{ref}=0.1 MPa).

Chemical Reactions Modelling

As customary (Chemkin, 2006, chap. 2), a general approach is used to model the kth chemical reaction involving S_k dir species that are the in-constituents and $S_{k,}$rev out-constituents. One obtains the reaction rate of species j when assuming that they participate in reactions k, all within V_j.

The reaction rate is given by:

$$q_k \equiv K_k^{dir} \prod_{m=1,\ldots,S_{k,dir}} [X]_{m,k}^{\beta_{m,in}^k} - K_k^{rev} \prod_{m=1,\ldots,S_{k,rev}} [X]_{m,k}^{\beta_{m,out}^k} \tag{15}$$

where the molar fractions is given by:

$$[X]_j = \frac{N_j}{V_j} \tag{16}$$

Coefficients β_m^k are usually the same as χ_{mk} (Chemkin, 2003).

The combined Arrhenius and Landau–Teller approach is taken as a default expression for the direct reaction kinetic constant K_{dir}^k (Jelezniak & Jelezniak, 2009). It is then given by:

$$K_k^{dir}(T) = A_k T^{a_k} \exp\left(\frac{-E_k}{RT}\right) \tag{17}$$

where constants Ak, ak and Ek are the parameters that are available in chemical kinetics databases (NIST, 2000). The reverse reaction kinetics constants are better measured through the equilibrium reaction rates constant K_k^{eq}, by relating them with the direct reaction constants, as they are linked through:

$$K_k^{rev}(T) = \frac{K_k^{dir}(T)}{K_k^{eq}(T)} \tag{18}$$

Any reaction is supposed to take place during a time τk, and generates during it an energy rate:

$$\dot{Q}_k = V_k q_k \Delta h_k \tag{19}$$

where Δh_k denotes, the enthalpy of reaction.

Phase Equilibria Modelling of Pure Substances

Vapour–liquid equilibrium condition can be written, using the definition of the Gibbs energy, as:

$$\mu_f = G_f = H_f - TS_f = H_g - TS_g = G_g = \mu_g \tag{20}$$

Rearrangement gives the relation between the entropy and the enthalpy of a phase transition:

$$\Delta S = S_g - S_f = \frac{1}{T}\left(H_g - H_f\right) = \frac{\Delta H}{T} \tag{21}$$

A small change of the chemical potential of one phase must be matched by the change of the chemical potential of the other phase. Therefore it concluded that:

$$d\mu_f = -S_f dT + V_f dp = -S_g dT + V_g dp = d\mu_g \tag{22}$$

Rearranging yields the following differential equation for the phase boundary curve:

$$\frac{dp}{dT} = \frac{S_f - S_g}{V_f - V_g} \tag{23}$$

Eq. (22) can be used to replace the transition entropy, resulting in Clapeyron's equation

$$\frac{dp}{dT} = \frac{\Delta H}{T \Delta V} \tag{24}$$

It is justified to assume $\Delta V \approx V_g = RT/p$; substitution into Eq. (24) and integration then yields the famous equation of Clausius and Clapeyron for the vapour pressure curve:

$$\ln \frac{p}{p_0} = -\frac{\Delta H}{R}\left(\frac{1}{T} - \frac{1}{T_0}\right) \tag{25}$$

Here (p_0, T_0) is a reference state on the vapour pressure curve.

The criteria for phase equilibrium Eq. (21) can be alternatively evaluated as follows:

$$\Delta G_{fg} = G_g - G_f = A_g^r - A_f^r - \ln\left(\frac{V_g}{V_f}\right) + \frac{p_{sat}}{RT}(V_g - V_f) \tag{26}$$

where p_{sat} denotes the saturation vapour pressure.

The equilibrium criterion can be rearranged to yield

$$\int_{V_f}^{V_g} p(V,T)\, dV = p_{sat}(V_g - V_f) \tag{27}$$

Eqs. (26) and (27) constitute a nonlinear equation for the vapour pressure, known as Maxwell's criterion.

The Maxwell equal area criterion is based on the equality of pressure, according to which a horizontal (constant pressure) line segment connecting the points at Vf and Vg on the pressure–volume

(PV) isotherm cuts off equal areas from the curve above and below. It can be used to compute vapour pressure curves from equations of state, provided that an initial guess for the vapour pressure is available for which the equation of state gives distinct liquid and vapour densities.

As shown in Fig. 1 Maxwell's criterion is evaluated as an objective function in which the initial pressure increases or decreases, depending on the curvature of the equation of state. The two-phase regions are located by resolving the Peng–Robinson equation of state with Cardano's formula. Maxwell's criterion is solved by means of the Marquardt–Levenberg method, from which the optimal pressure and molar volume at saturation points are obtained.

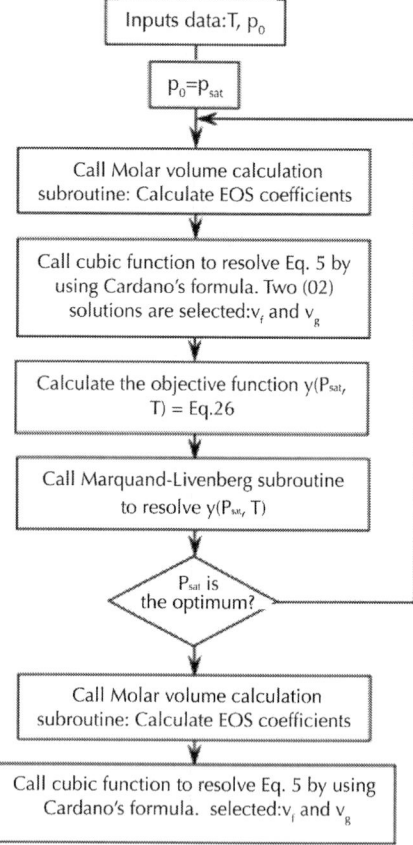

Figure 1: General diagram of vapour–liquid equilibrium (VLE) calculation.

A SUITABLE APPROACH FOR SAFETY ANALYSIS

As indicated above, safety analysis usually involves two parts. The deterministic analysis which aims to evaluate the transient behaviour of chemically reacting system can be coupled with the probabilistic analysis with the goal of obtaining an estimation of the exceedance frequencies of specified safety limits. To illustrate the safety analysis approach, we present in Fig. 2 an adequate dynamic safety analysis model based on thermal-hydraulic analysis. No attempt is made to develop the probabilistic safety analysis part rather to show the importance which plays the deterministic safety analysis in the performance of the probabilistic safety analysis.

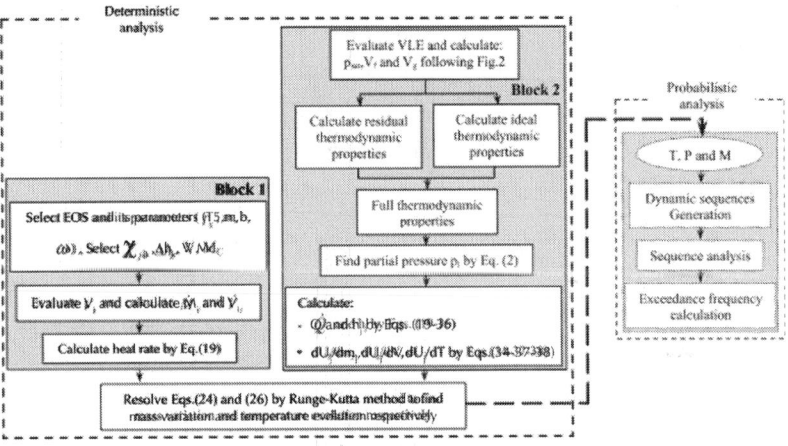

Figure 2: General methodology for dynamic safety analysis.

As depicted in Fig. 2, the methodology has two main steps:

Deterministic Analysis Methodology

Deterministic analysis is based on thermal-hydraulic analysis, used to compute the evolution of pressure, temperature, molar concentrations of gaseous mixtures at different rates during the severe accident, with exothermic reactions.

As shown in Fig. 2, at the beginning we select the cubic equation of state with its parameters. For each reaction, we introduce the stoichiometric coefficients and the enthalpy of reaction. We also introduce the total volume of the containment or tank with the heat capacity of the solid structures.

Block 1 describes the main steps to be followed to resolve mass transfer equation for different species. Block 2 describes the main steps through which the temperature equation with different time steps is resolved.

Mole (Mass) Variation Calculation (Block 1)

In addition to the accounting for the sources of species as a result of the chemical reactions, mass transport is important when there is material transfer across the volume surface, its simpler case being a set of known incoming mixture mole flows as a function of time, with a volumetric mixture flow jr at surface piecer. Then, a typical mole conservation equation is of the form:

$$\frac{dN_j}{dt} = \sum_r \left[\frac{\varepsilon_{jr} \rho_j^{in} j_r^{in}}{M_j} - \frac{\varepsilon_{jr} N_j}{V} j_r^{out} \right] + \sum_k \chi_{jk} q_k(t) \theta(t - \tau_{star\ k})$$
$$\theta(\tau_{start\ k} + \Delta_k - t) \tag{28}$$

where τ_{start} k; Δ_k are the start and duration of reaction k. $\varepsilon_{jr}=1,0$ just indicates whether or not species j goes through surface piece r.

As shown in Block 1 and at the beginning, the specific volume vj is evaluated. After, we establish a loop for each gas to calculate the mass flow $M_j = \varepsilon_{jr} \rho^{in} j^{in}$ or the volume flow $V_j = \varepsilon_{jr} j^{out}_r$. At last, when we have completed the loop for each gas, we calculate the heat that appears in the chemicals reacting using Eq.(19). The temporal evolution of the different species formed by different reactions inside the containment is calculated via Eq. (28). It is solved through a modified Runge–Kutta numerical method using a smaller time step than the input-predetermined to account for the faster speed of mass transfer versus heat capacity-mediated energy transfer.

Temperature Evolution Calculation (Block 2)

The first law of thermodynamic for the variation of the overall internal energy in our case will be imposed through:

$$\frac{dU}{dt} = \dot{Q}(t) + \sum_j (\dot{m}h_{in})_j - P\frac{dV}{dt} \tag{29}$$

where $\dot{Q}(t)$ is the calorific power exchange with the outside and $\sum_j (\dot{m}h_{in})_j$ is the calorific power inlet due to the gases that are entering. We obtain a differential equation for the temperature, via

$$\frac{dU}{dt} = \left.\frac{dU}{dt}\right|_{structure} + \left.\frac{dU}{dt}\right|_{mixture} \tag{30}$$

Also, we know that:

$$\left.\frac{dU}{dt}\right|_{structure} = M_C \frac{dT}{dt} \tag{31}$$

And

$$\left.\frac{dU}{dt}\right|_{mixture} = \sum_j \left(\frac{dU_j}{dV}\frac{dV}{dt} + \frac{dU_j}{dm_j}\frac{dm_j}{dt} + \frac{dU_j}{dT}\frac{dT}{dt}\right) \tag{32}$$

where M_C is the heat capacity of the solid structures included in the volume. If we resolve for dT/dt, we obtain the following temperature differential equation:

$$\frac{dT}{dt} = \frac{\dot{Q} + \sum_j (\dot{m} h_{in})_j - \sum_j \left[\frac{\partial U_j}{\partial V} + p_j\right] dV/dt - \sum_j \frac{\partial U_j}{\partial m_j} \frac{dm_j}{dt}}{M_C + \sum_j \frac{\partial U_j}{\partial T}}$$

(33)

The analysis of the different terms that appear in the temporary variation of the temperature gives the following equations:

- Pressure p

Pressure p_j is the partial pressure of each component in the mixture. It can be calculated using the Eq. (2).

- **Variation of internal energy with respect to mass**

For each component j present in the mixture:

$$\frac{\partial U_j}{\partial m_j} = u_j$$

(34)

where:

$$u_j = h_j - p v_j$$

(35)

h_j is a sum of residual and ideal terms where:

$$h_j = h_j^r + h_j^{id}$$

(36)

h_j^r and h_j^{id} are obtained by Eqs. (8) and (14) respectively.

- **Variation of internal energy with respect to the volume**

For each component present in the mixture:

$$\frac{\partial U_j}{\partial V} = T\left(\frac{\partial p_j}{\partial T}\right) - p_j \qquad (37)$$

- **Variation of internal energy with respect to temperature**
  For each component j presents in the mixture:

$$\frac{\partial U_j}{\partial T} = C_{v,j} \qquad (38)$$

$C_{v,j}$ is the sum of residual and ideal terms where:

$$C_{v,j} = C_{v,j}^r + C_{v,j}^{id} \qquad (39)$$

$C_{v,j}^r$ and $C_{v,j}^{id}$ are obtained using Eqs. (9) and (13) respectively.

As indicated below, Block 2 solves the temperature equation. All terms of Eq. (33) are calculated using their corresponding Eqs. (2), (19), (34), (36), (37) and (38). Finally, with these values, we resolve the differential equation Eq. (33) using a modified Runge–Kutta numerical method.

Probabilistic Analysis Methodology

While deterministic analyses may be used to verify that acceptance criteria of protection design are met, probabilistic safety analyses may be used to determine the probability of damage for each protection barrier. Probabilistic safety analysis may thus be a suitable tool to compute the exceeding frequency for the sequences created from a common initiating event. As shown in Fig. 2, the global simplified diagram for probabilistic analysis methodology computes the exceeding frequency of sequences created from the evolution of process variables (temperature, pressure and mass) given by the deterministic analysis. The computing process consists on the following steps:

- Dynamic sequences generation: the objective is to generate the dynamic event trees (DET) through all possible scenarios, delineates each sequence of DET stemming from an initiating event. For each sequence, the corresponding times and probabilities are provided.
- Sequence analysis: aims to identify the damage domain of all paths which constitute one sequence. Each sequence is characterized by a set of times and a set of sensitive parameters, such that for certain combinations of them damage affected to each path may be generated. Thus, the damage domain is a volume in the multi-dimension space of times and parameters, each point in it representing a different transient. Each point should be evaluated by using the information of process variables evolution given by the deterministic analysis.
- Risk assessment: Once the damage domain has been identified, the analysis may proceed with the calculation of the exceeding frequency of the contribution of all paths within the sequence. Each transient belonging to the damage domain is revisited in order to identify the information needed for the calculation of the frequency density. Numerical integration of the exceeding frequency will then consist in computing the Q-kernel on the damage points inside the sequence sampling domain by summing up all the contributions.

RESULTS AND ANALYSIS

General Assumptions

Safeguards Available

Protection system is designed and built to prevent any release of radioactive and toxic products, in order to avoid any consequence on the population and environment. Table 1 shows some of the similarities between the safeguard systems provided for the Bhopal tank (Gupta, 2002) and the nuclear containment (Peeters, 2007).

Table 1: Safeguards available

Nuclear containment	Tank 610
Containment filtered venting system can be used to release steam in order to depressurize the containment and terminate the possible release of radioactive substances into the environment.	Vent gas scrubber system for neutralizing the toxic release material released from various equipment of MIC plant. This system is connected to flare tower system for burning the vent gases at a controlled rate.
Containment spray system can be operated to prevent any temperature and pressure rises beyond the designed limits.	Water spray system (or deluge system) for neutralizing MIC if unreacted MIC escapes even after scrubbing.
Containment heat removal system which provides cooling of the in-containment refuelling water storage tank. Consequently, pressure and temperature decrease.	Refrigeration system to keep the storage tank material below 5 °C.
Pressure relief valve system to decrease the pressure inside the containment	Relief valve system connected to the vent system from which the MIC was released after rupture of the safety disc.

Nuclear Containment H_2 Issue

As earlier described, the hydrogen issue is one of the major risks challenging the containment integrity of a nuclear power plant under severe accidents. In our simulations we try to reproduce its basic features. The containment is assumed to be homogeneous, composed by H_2, O_2, N_2 and a water/steam mixture. An injection of a given volumetric H_2 flow at high temperature for a short time is the initiator. The initial conditions of the containment are a mixture of air and steam at equilibrium at the initial pressure and temperature. The mixture reacts with the incoming flow leading to a change in the relative fractions of the constituents. We assume that as soon as flammability conditions are reached, combustion takes place, subsequent several such events being possible.

Bhopal

As mentioned, there are many reactions which tend to be exothermic and vigorous. From a practical standpoint, and to simplify our problem only a limited number of those reactions that conceivably could have taken place are considered in our study. We suppose that only methyl-isocyanate reacts exothermically with water to give methylamine and carbon dioxide. The methyl-isocyanate system constitutes a pure mixture of steam-liquid phase. The methylamine and CO_2 formed the incondensable gaseous mixture region. For our simulation, it is supposed also that the tank 610, contains only a gaseous mixture of methyl-isocyanate and steam water with specific volume for each one being v>vg.

Nuclear Containment Case

Validation of the H_2 Combustion

The hydrogen issue is one of the major risks challenging the containment integrity of a nuclear power plant. Indeed, during severe accidents, a large amount of hydrogen is produced.

However, hydrogen is a very flammable gas. So, as soon as an igniter appears, combustion occurs. This combustion leads to an abrupt increase of the pressure and temperature inside the containment. The containment mass is assumed to be homogeneous, composed of H_2, O_2, N_2 and steam water mixture.

The steam in the reactor reacts with the steel and zirconium. These chemical reactions lead to the production of hydrogen. The initial values of the process variables are supposed equal to: T_{ini} = 373.15 K, mass of H_2 = 1000 kg, mass of O_2 = 12,000 kg, mass of steam H_2O = 2500 kg, mass of N_2 = 17,000 kg.

Residual thermodynamic properties and partial pressure of each component are calculated with the Peng–Robinson equation of state. The user can select other equations of state like Van der Waals, Redlich–Kwong, Soave–Redlich–Kwong, etc.

- **Mass variation inside the containment**

Fig. 3 shows the mass variation of the steam water, oxygen, hydrogen and nitrogen inside the containment.

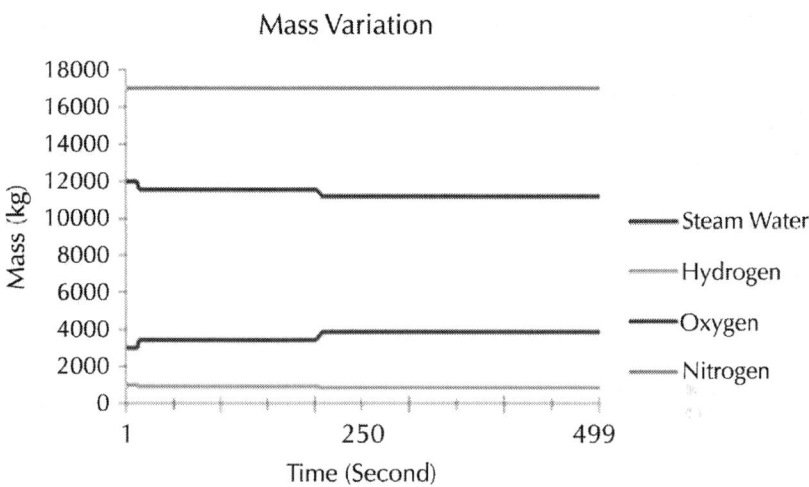

Figure 3: Mass variation of steam water, oxygen, hydrogen and nitrogen.

At the beginning of the first combustion at 10 s, the mass of reactants composed of the oxygen and the hydrogen decrease gradually. In addition, the mass of product composed by the steam water increase. At 200 s and during 10 s of the second combustion the mass of product increase and the mass of reactants decrease. At the end of each reaction, the mass of each component is maintained as constant and the containment mixture evolves to its steady state.

- **Temperature and pressure evolution inside the containment**

Fig. 4 and Fig. 5 show the temperature and the pressure evolution inside the containment. At 10 s, the first combustion of hydrogen takes place. We can observe that the temperature and the pressure increase. When the first combustion has ended at 15 s of the simulation, the containment mixture evolves to its steady state. The thermodynamic variables remain constant until 200 s when the second combustion occurred. The temperature and the pressure increase again. When the first combustion has ended after 10 s of the simulation, the containment mixture evolves to its steady state.

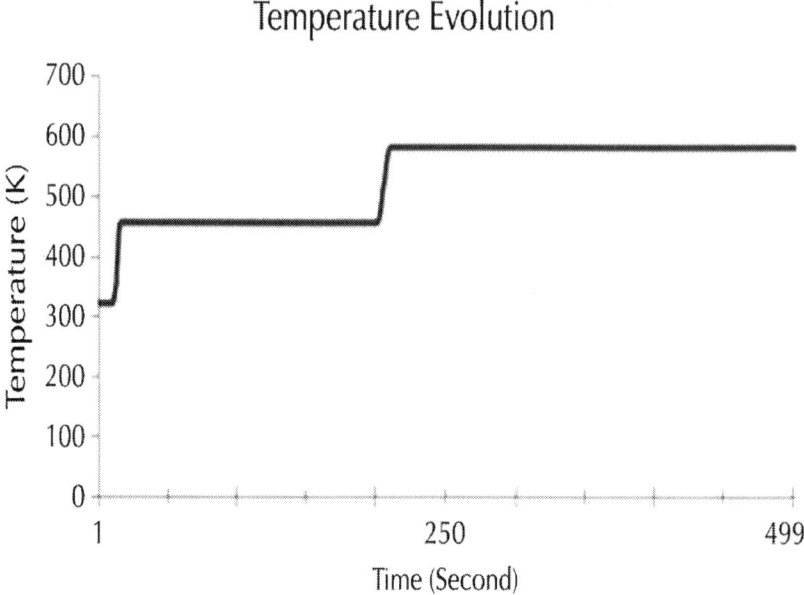

Figure 4: Temperature evolution inside the containment.

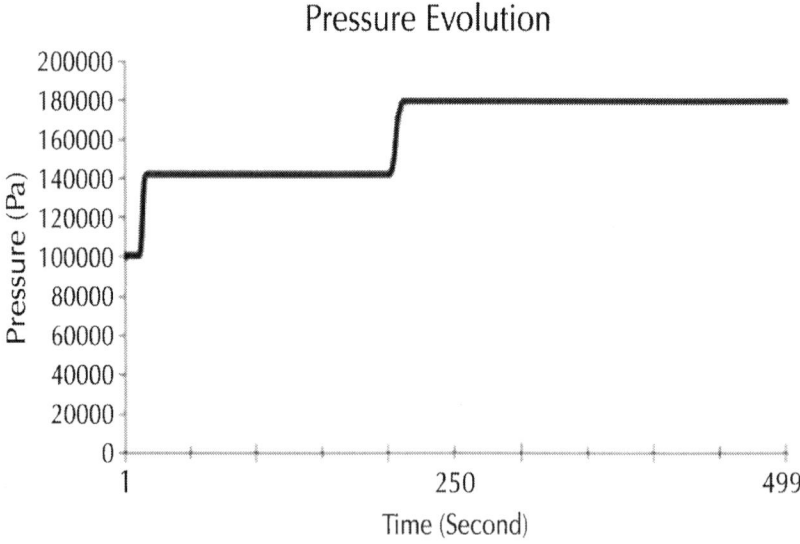

Figure 5: Pressure evolution inside the containment.

Validation of the Containment Heat Removal System (CHRS)

This system is one of the safeguard systems of the nuclear power plant and consists of the pulverization of cold water droplets in the containment. As the droplets temperature is lower than the atmosphere temperature, the steam of the atmosphere condenses on these droplets. Consequently, the temperature and pressure inside the containment decrease. We will introduce a quantity of cold water droplets during 100 s by the addition of another pipe inside the containment.

Fig. 6 and Fig. 7, show the regulation of the temperature and the pressure in the case of one injection of droplets at 25 °C with flow rate equal to 0.2 kg/s. Injection ended, we obtain the steady state directly, even though, the temperature and the pressure will rise until getting the steady state when the injection is finished.

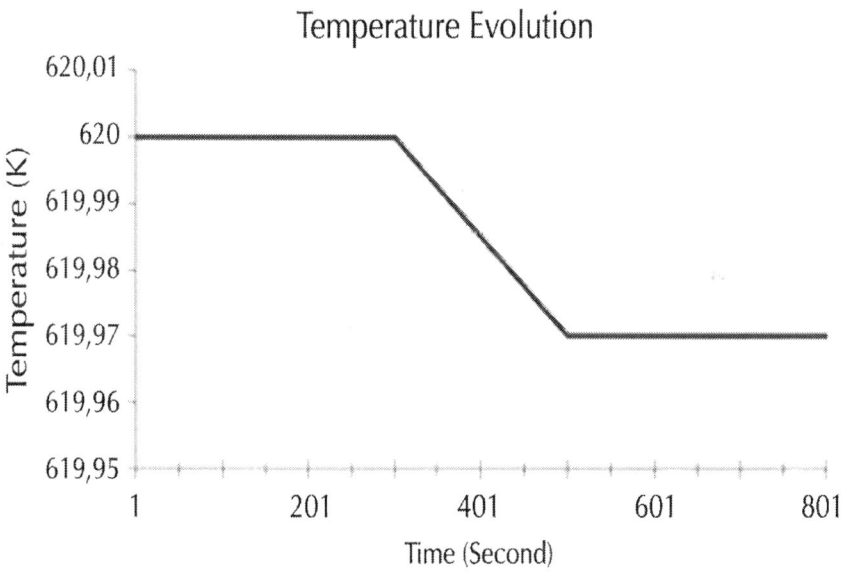

Figure 6: Temperature evolution inside the containment.

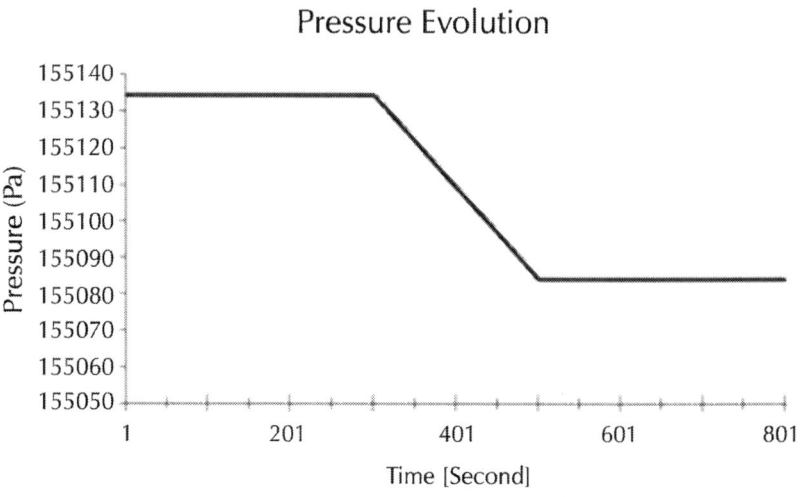

Figure 7: Pressure evolution inside the containment.

Validation of the Relief Valve

For this case, we have taken into account the addition of another pipe. With this pipe, we have implemented the relief valve. The objective of the relief valve is to limit the pressure of the system during the variations of power when pressure reaches the set point. Another objective of this valve is to relieve the pressure during an accident.

During the simulation, it has considered that this valve is in the closed state. When pressure inside the containment reaches the set point, the valve discharges only steam water.

For the validation of the performance of this valve, we will suppose, three quantities of mass flow during the same time. The first one case with a mass flow lower, the second one case with a medium mass flow and the last one case with the maximum mass flow. We can see in the next figures the three types of mass flow.

In Fig. 8 and Fig. 9, we can see the evolution with respect of time of two variables of the containment like the total pressure, the temperature. We show the containment in steady state, then we do the extraction and finally, we allow again the containment to evolve to the steady state.

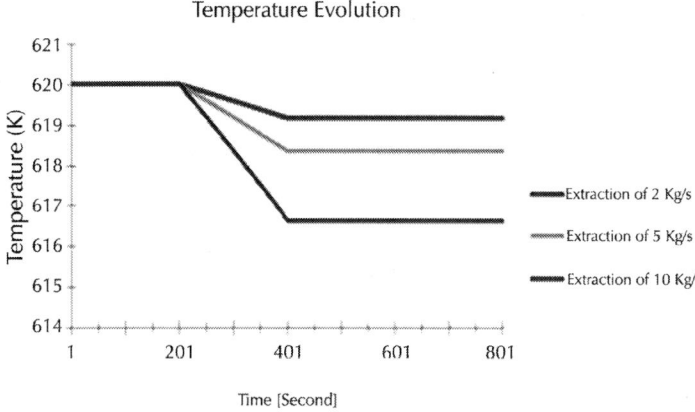

Figure 8: Temperature evolution inside the containment.

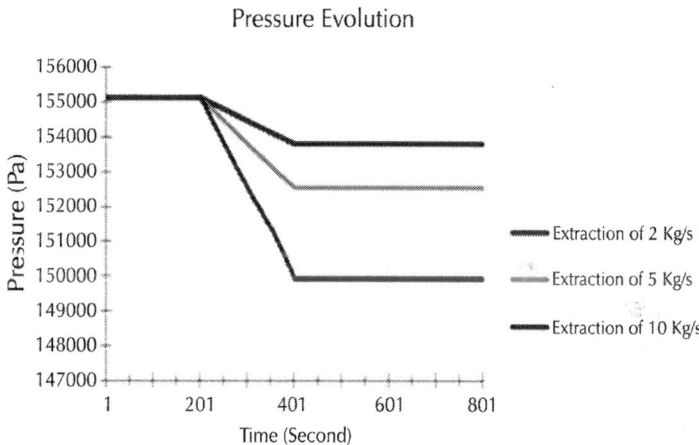

Figure 9: Pressure evolution inside the containment.

As we can see for the three cases, the relief valve works correctly. In the three cases the pressure decreases as well as the temperature in the containment decrease. As it is logical, when we extract more quantity of steam, the higher is the decrease.

As we could see in the other inputs, when the extraction of steam has ended, the steady state is gotten directly when we stop doing the extraction.

Bhopal Tank 610 Case

Validation of the MIC Entrance

The direct cause of the chemical reactions between water and methylisocyanate was the entrance of water inside the tank. Many theories have been advanced on how water might have entered the methylisocyanate storage tank. Water could have entered at some point in the nitrogen line near the tank, water could have entered through the refrigeration line, or water could have entered directly through the process vent system.

For each theory we suppose that the MIC entered in the tank with different mass flows, we suppose also, that the initial temperature inside the tank at the beginning of the reactions is 298 K. Depending on the initial pressure, we will simulate an accidental case with assumption that the tank contains an initial steam of water of 6000 kg at 10^5 Pa.

As described in Fig. 2, for the case of this example, the modeller changes only the input file which contains the initial temperature inside the tank, stoichiometric coefficient, molar mass, enthalpy of reaction. Depending on the equation of state, the input file contains also the, co-volume parameter, acentric factor and specific temperature for each component of the mixture.

- **Mass variation inside the tank**

Very similar to the case of nuclear containment, Fig. 10 and Fig. 11 show the mass variation of the steam water, methyl-isocyanate, Methylamine and carbon dioxide inside the tank.

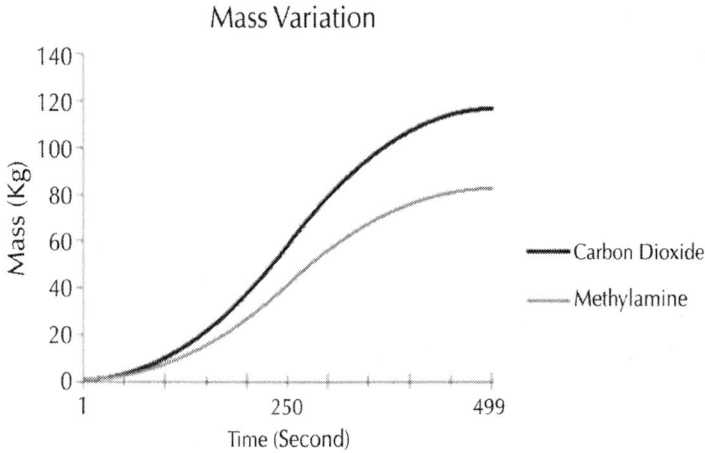

Figure 10: Mass variation of methylamine and carbon dioxide.

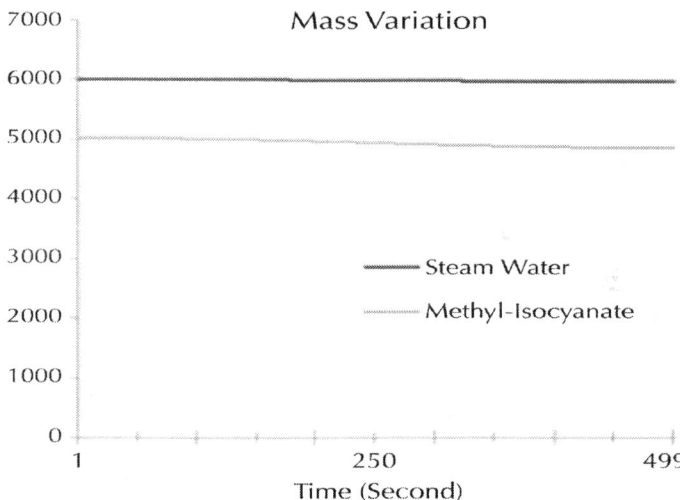

Figure 11: Mass variation of steam and methyl-isocyanate.

At the beginning of the reaction between steam water, and methyl-isocyanate at 0 s, the mass of reactants composed of methyl-isocyanate and steam water decrease gradually. In addition, the mass of product composed of the Methylamine and the CO_2 decrease. During simulation the mass of product increases and the mass of reactants decreases.

The thermodynamic variables continue to vary and the tank mixture evolves to its unsteady state.

- **Temperature and pressure evolution inside the tank**

Very similar to the case of nuclear containment, Fig. 12 and Fig. 13 show the temperature and the pressure evolution inside the tank. At 0 s, the exothermic reaction between steam water and methyl-isocyanate takes place. We can observe that the temperature and the pressure increase. During the simulation, the thermodynamic variables continue to vary and the tank mixture evolves to its unsteady state.

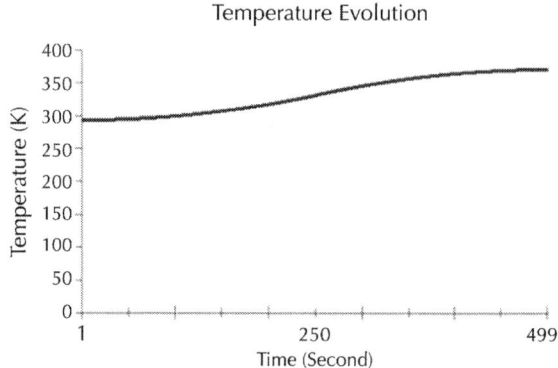

Figure 12: Temperature evolution inside the tank.

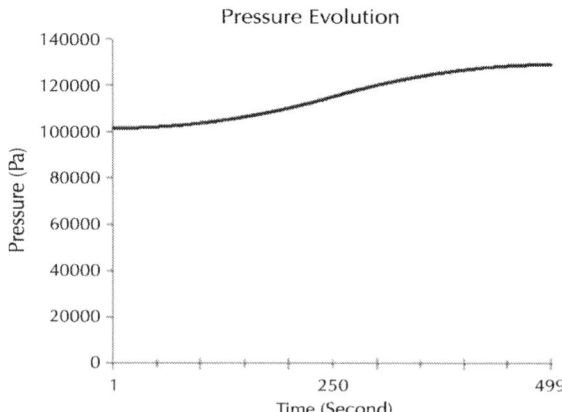

Figure 13: Pressure evolution inside the tank.

Validation of the Refrigeration System

The overall task of the refrigeration system is to keep the tank contents below 5 °C, by the injection of Freon through circulating cooling pipe. Very similar to the nuclear case, we will introduce a quantity of Freon during 100 s by the addition of another pipe connected to the tank.

Fig. 14 and Fig. 15, show the regulation of the temperature and the pressure in the case of injection of Freon at boiling temperature with a flow rate equal to 0.1 kg/s. Injection ended, we obtain the steady state directly, even though, the temperature and the pressure will rise until getting the steady state when injection is finished.

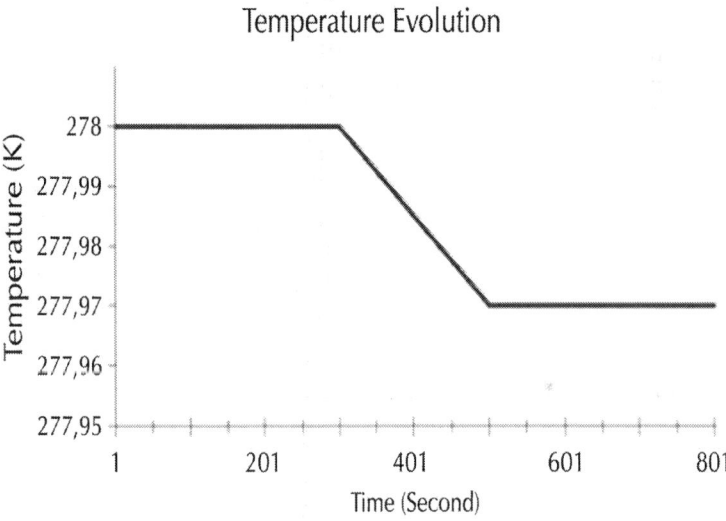

Figure 14: Temperature evolution inside the tank.

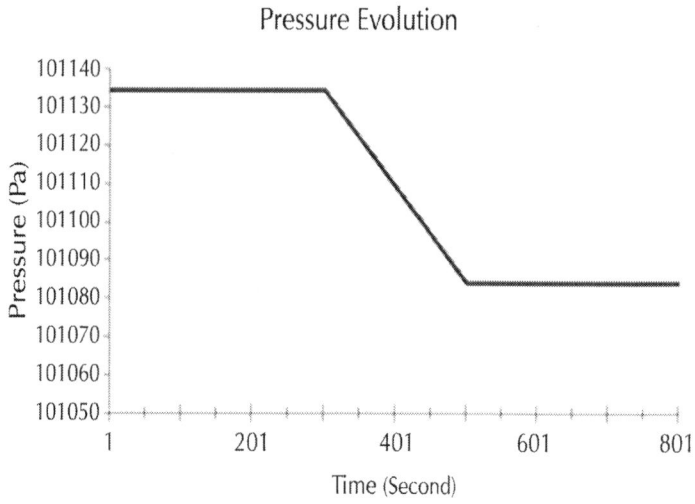

Figure 15: Pressure evolution inside the tank.

Validation of the Relief Valve

Very similar to the case of nuclear containment, we have implemented into the pipe configuration a relief valve with the objective being to limit the pressure of the system during the variations of power.

During the simulation, it has been considered that this valve is in the closed state. When pressure inside the tank reaches the set point, the valve discharges only steam of MIC.

For the validation of using this valve, we will suppose, three quantities of mass flow always during the same time. The first one case is with a lower mass flow, the second case is with a medium mass flow and the last case is with a maximum mass flow. We can see in the next figure the three types of mass flow.

As shown in Fig. 16 and Fig. 17, we can see the evolution with respect of time of three variables of the tank like the total pressure, the temperature. As we see, the dynamic of the extraction is very similar to the case of nuclear containment in which, first we show the tank in steady state, then we do the extraction and finally, we allow again that the tank condition evolves to the steady state.

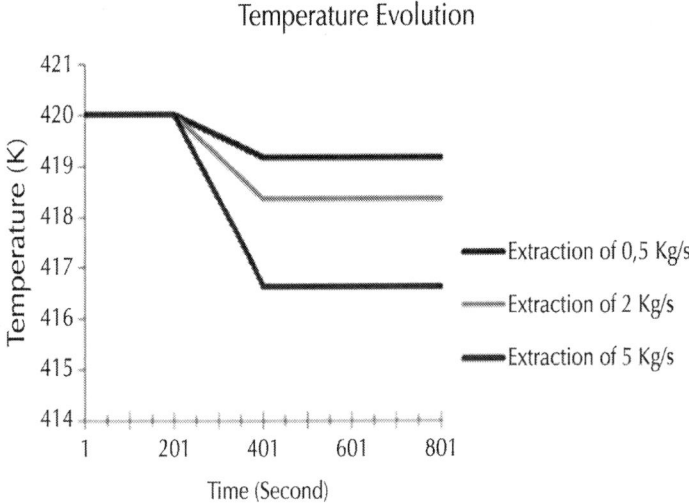

Figure 16: Temperature evolution inside the tank.

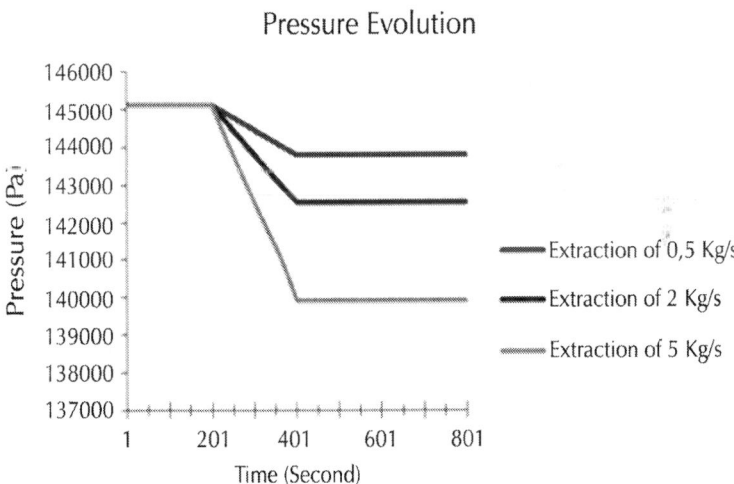

Figure 17: Pressure evolution inside the tank.

As we can see for the three cases, the relief valve works correctly. In the three cases the pressure decreases as well as the temperature in the tank decreases. As it is logical, when we extract more quantity of steam of MIC, the higher is the decrease.

As we could see in the other inputs, when the extraction of steam of MIC has ended, the steady state is achieved directly when we stop doing the extraction.

CONCLUSIONS

We have developed an adequate tool to transfer some features of thermal-hydraulic modelling from the nuclear industry to the simulation of similar scenarios in chemically reacting facilities, something necessary when the risk related decision making requires the performance of a large number of scoping deterministic analyses.

To verify its feasibility, the tool has been tested in a reference case for both the nuclear and chemical/petrochemical industries taking into account the possibility of future developments. Our application to study how the simplified chemical reactions and system configurations could impact on Bhopal like disaster and H_2 deflagrations in nuclear containments verifies a significant analysis power that can be adapted to different purposes in the industry. Future applications to be performed with our tool will introduce binary mixture equilibrium as well as two-phase non-equilibrium transient conditions. They will use the gas–liquid–solid transitions to increase the computing power. Some modules currently under development will provide the possibility for introducing binary mixtures to compute thermodynamic properties by using different equations of state for all phases.

Because of the relation of deterministic and probabilistic safety approaches, future developments will be connected to both. A practical specific module with the objective of assessing the safety space will be pursued. The safety space in the context of probabilistic safety can be understood as an extension of the PSA event trees and the uncertainty analysis methods, aimed at obtaining an estimation of the exceedance frequencies of specified safety limits.

REFERENCES

1. Berger, P., El Tahhann, R., Moulin, G., & Viennot, M. (2003). High temperature oxidation of zirconium and zircaloy-4 under applied load: nuclear microprobe study of the growth of the oxide.

2. Breitung, W., & Royl, P. (2000). Procedure and tools for deterministic analysis and control of hydrogen behavior in severe accidents. Nuclear Engineering and Design, 202, 249e268, Elsevier.
3. Chemkin Collection. (June 2006). Thermodynamics and chemical rates expressions.
4. Chemkin Collection. (2003). Gas-phase kinetics core utility manual.
5. Eckerman, I. (2005). The Bhopal gas leak: analyses of causes and consequences by three different models. Journal of Loss Prevention in the Process Industries, 18, 213e217, Elsevier.
6. Gupta, J. P. (2002). The Bhopal gas tragedy: could it have happened in a developed country? Journal of Loss Prevention in the Process Industries, 15, 1e4, Elsevier.
7. Hasan, O., Stanley, S., & Arvind, V. (1998). Modeling vaporeliquid equilibria: Cubic equations of state and their mixing rules. Cambridge University Press.
8. Jelezniak, M., & Jelezniak, I. (2009). Differential equations of gas-phase chemical kinetics. Chemked e a program for chemical kinetics of gas-phase reactions. http://chemked.com.
9. Kirchsteiger, C. (1999). On the use of probabilistic and deterministic methods in risk analysis. Journal of Loss Prevention in the Process Industries, 12, 399e419, Elsevier.
10. Labeau, P. E., Smidts, C., & Swaminathan, S. (2000). Dynamic reliability: towards an integrated platform for probabilistic risk assessment. Reliability Engineering and System Safety, 68, 219e254, Elsevier.
11. Lepkowski, W. (1985). Bhopal, Indian city begins to heal but conflicts remain. Chemical & Engineering News, 63(48), 18e34, ACS Publications.
12. Medina, H., Arnaldos, J., & Casal, J. (2009). Process design optimization and risk analysis. Journal of Loss Prevention in the Process Industries, 22, 566e573, Elsevier.
13. NIST Chemical Kinetics Database on the Web. (2000). A compilation of kinetics data on gas-phase reactions. http://kinetics.nist.gov/kinetics/index.

14. NIST chemistry web book. (2008). http://webbook.nist.gov.
15. Peeters, A. (2007). Application of the stimulus-driven theory of probabilistic dynamics to the hydrogen issue in level-2 PSA. PhD thesis, Université Libre de Bruxelles.
16. Themistocles, D., Anibal, L., Russell, L., Sureerat, S., & Chan, J. K. (1986). Studies of methyl isocyanate chemistry in the Bhopal incident. The Journal of Organic Chemistry, 51, 3781e3788, ACS Publications.
17. Tregoures, N., Philippot, M., Foucher, L., Guillard, G., & Fleurot, J. (2010). Reactor cooling systems thermal-hydraulic assessment of the ASTEC V1.3 code in support of the French IRSN PSA-2 on the 1300 MWe PWRs. Nuclear Engineering and Design, 240, 1468e1486, Elsevier.
18. Vidal, J. (1997). Thermodynamique, Application au génie chimique et à l'industrie pétrolière. Edition Technip.

Chapter 8

Conventional and Dynamic Safety Analysis: Comparison on a Chemical Batch Reactor

L. Podofillini and V.N. Dang

Paul Scherrer Institute, Switzerland

ABSTRACT

Dynamic safety analysis methodologies are an attractive approach to tackle systems with complex dynamics (i.e. with behavior highly dependent on the values of the process parameters): this is often the case in various areas of the chemical industry. The present paper compares analyses with Probabilistic Safety Assessment (PSA)/Quantitative Risk Assessment (QRA) methods with those from a dynamic methodology (Monte Carlo simulation). The results of a case study for a chemical batch reactor from the literature, overall risk figure and main contributors, are examined. The comparison has shown that, provided that the event success criteria are appropriately defined, consistent results can be obtained; otherwise important accident scenarios, identifiable by the dynamic Monte Carlo simulation, are possibly missed in the application of conventional methods. Defining such criteria was quite resource-intensive: for the analysis of this small system, the success criteria definitions required many system simulation runs (about 1000). Such large numbers of runs may not be practical in industrial-scale applications. It is shown that success criteria obtained with fewer simulation runs could have led to different quantitative PSA results and to the omission of important accident scenario variants.

INTRODUCTION

The risk models in conventional Probabilistic Safety Assessment (PSA) (or Quantitative Risk Assessment, QRA, as typically referred to in the chemical industry) are based on logical analytical techniques, e.g. event and fault trees. These provide a structured framework to represent accident scenarios and system failure and success mechanisms. The accident sequence models and the system success criteria are developed based on a set of analyses (typically thermo-hydraulic in the nuclear industry and process dynamics in the chemical one), addressing bounding scenarios and success criteria. The selection of this set typically requires a significant effort. Challenges arise when differences in the sequential order or in the timing of the success and failure events along an accident scenario strongly affects the subsequent accident scenario evolution, and when the evolution of the process variables (temperatures, pressures, mass flows, etc.) affects the occurrence probabilities of the events (and thus the subsequent scenario evolution) [1].

To cope with these issues, dynamic methodologies integrate plant physical models (typically thermo-hydraulic codes in the nuclear power industry) and operating crew models into stochastic simulation engines. The basic idea is to directly generate the accident scenarios by simulating the interactions of the plant systems, components, and operating crew over time from the initiating events. These scenario variants are automatically generated by the stochastic simulation engine (within certain specified boundaries), without the need for the analyst to specify the system state each time a new variant is explored. Reviews of the available techniques can be found in Refs. [1], [2],

[3], [4], [5], [6], [7] and [8]. Most relevant applications of dynamic methodologies focused on nuclear power plants, addressing modeling of the operating crew response [8], [9], [10], [11] and [12], of digital instrumentation and control systems[13] and [14], and level 2 and 3 PSA [6].

The difficulty of comprehensively modeling accident scenario variants within the typical event/fault tree framework is recognized in various areas of the chemical industry as well, especially when the process dynamics is complex and highly dependent on the values of the process parameters [15], [16], [17], [18],[19] and [20]. Yet, the use of integrated physical-stochastic models, which automatically generate scenario variants, is still rare [16] and [17]. A direct consequence of the mentioned difficulties is that the results of conventional PSA/QRA are strongly dependent on the accident sequence models and event success criteria, as discussed in Refs. [15] and [21] with application to the chemical industry.

To contribute to the understanding of the possible challenges of the traditional framework and of the potentials of the dynamic approach, the present paper presents a comparison of conventional and dynamic PSA (with Monte Carlo simulation). The case study for the comparison is a chemical batch reactor from the literature [22]. In the analyzed system, the scenario evolution is highly sensitive to the timing of operator actions and component failures and on the values of the process variables at the time these events occur. This makes the example particularly suitable for a dynamic methodology. The focus of the comparison is on understanding how success criteria influence the conventional PSA results. Note a thorough comparison of the traditional and dynamic approaches has been attempted in some publications[1] and [13], but not with the specific focus on the success criteria definition.

The paper is organized as follows. The next Section 2 describes the system under analysis, under nominal and accident conditions. Section 3 presents the safety analysis carried out with conventional PSA. Two hypothetical analysts are introduced (analyst A and B), each taking a different bounding scenario for the determination of the accident sequence and success criteria. The results from the two hypothetical analysts are compared, showing how the selection of the reference scenario influences the PSA results.Section 4 presents the Monte Carlo analysis of the same system. MC and PSA results are compared inSection 5, in terms of both the predicted risk figure and most important contributors to risk. Discussion and conclusions close the paper.

SYSTEM DESCRIPTION

The system under analysis is a chemical batch reactor in which a highly exotermic process takes place[22]. In the next Section 2.1, the chemical–physical model of the reactor is described with respect to its nominal, safe-operative condition. Deviations from the nominal behavior may arise due to stochastic failures of the system components, as described in Section 2.2.

Model of the Reactor in Nominal Conditions

The system considered is the chemical batch reactor of Fig. 1[22]. Two parallel, highly exothermic reactions are carried out in the reactor:

A+B→C

A+C→D

where A and B are the raw materials, C is the desired product, and D an undesirable by-product, respectively. The batch reactor dynamics is modeled by the following equations of material balances in the reactor [22]:

$$\frac{dM_A}{dt} = -k_1 M_A M_B - k_2 M_A M_C \tag{1}$$

$$\frac{dM_B}{dt} = -k_1 M_A M_B \tag{2}$$

$$\frac{dM_C}{dt} = +k_1 M_A M_B - k_2 M_A M_C \tag{3}$$

$$\frac{dM_D}{dt} = +k_2 M_A M_C \tag{4}$$

and energy balances:

$$\frac{dT_r}{dt} = \frac{Q_r + Q_j}{M_r Cp_r} \qquad (5)$$

$$\frac{dT_j}{dt} = \frac{F_j \rho_j Cp_j (T_{jsp} - T_j) - Q_j}{V_j \rho_j Cp_j} \qquad (6)$$

where M_i is the amount of moles of reagent "i"; k_1 and k_2 are the reaction rates, dependent on the reactor temperature T_r; T_j is the jacket temperature, with T_{jsp} being the setpoint value of the associated control system; Q_r and Q_j are the amount of heat produced by the chemical reaction and exchanged through the jacket-reactor interface, respectively; F_j is the mass flow in the reactor jacket. The meaning of the other variables and parameters are reported in Appendix A, together with their chemical and physical relations.

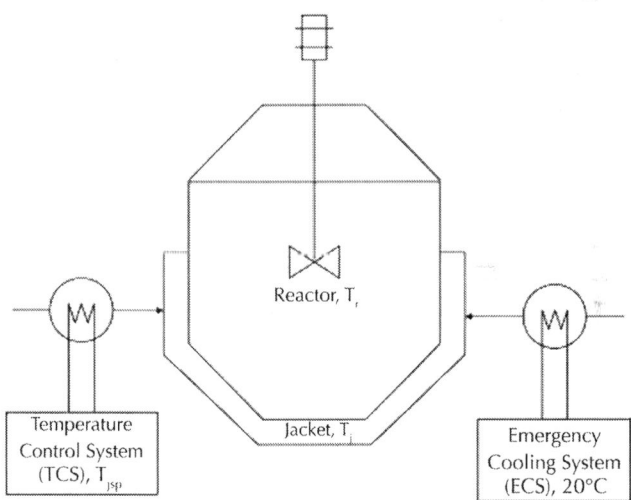

Figure 1: The batch reactor system.

The dynamics of the reaction is simulated by solving Eqs. (1), (2), (3), (4), (5) and (6) over a production batch of 240 min. In nominal operation, the initial conditions of MA, MB, MC and MD for each production batch are 12, 12, 0, and 0 kmol, respectively [22]. The initial values of both reactor and jacket temperatures are set to 20 °C. The jacket setpoint temperature Tjsp, controlled by the Temperature

Control System (TCS), is 71 °C and it is kept constant over the 240 min duration of each production batch. The TCS can be thought of as a heat sink at the constant temperature Tjsp. The heat is transferred from the TCS to the jacket (and viceversa) through the heat exchanger shown in Fig. 1. The values of the other process parameters used in the reactor model are listed in the Appendix A.

The behaviors of the reactor temperature Tr and of the concentration of the desired reaction product C over the batch duration are reported in Fig. 2. During the reactor operation, the reactor temperature increases up to about 90 °C (heat-up phase) and then decreases to the jacket setpoint temperature (Fig. 2, top), while the concentration of product C increases up to 6.4 kmol (Fig. 2, bottom). The cool down phase is critical: the heat produced by the exothermic reaction has to be removed. If the setpoint temperature control to 71 °C were not active, runaway conditions would arise and the reactor temperature would continue to increase, toward unsafe conditions.

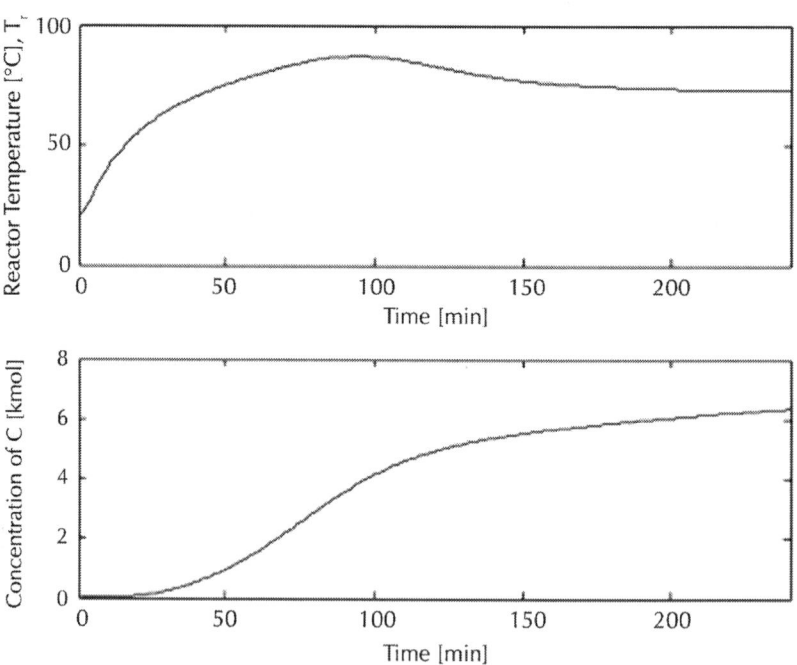

Figure 2: Behaviors of reactor temperature Tr and concentration of product C in nominal operative conditions.

Deviations from Nominal Conditions

Possible deviations in the process initial conditions and parameters (due for example to calibration errors) are modeled by Gaussian distributions applied to the values of the following process parameters: initial reagent concentrations, MA(0), MB(0), initial reactor temperature $T_r(0)$, mass flow in the reactor jacket, $F_j(0)$ and jacket setpoint temperature, T_{jsp}. Table 1 reports the characteristics of the Gaussian distributions. Given the exothermic character of the chemical reaction, such failures can lead to runaway conditions.

Table 1: Nominal values (means) and standard deviations of the process parameters

Variable	Nominal value (mean)	Standard deviation
A concentration, MA(0) [kmol]	12	1.0
B concentration, MB(0) [kmol]	12	1.0
Reactor temperature, Tr(0) [°C]	20	2.0
Jacket mass flow, Fj (0) [m3/min]	0.348	0.05
Jacket setpoint temperature, Tjsp [°C]	71	1.5

The effect of deviations in the process parameters (deviating one at a time) is shown in Fig. 3. For example, the bottom-left plot of Fig. 3 shows that the higher the deviation from the nominal value of the jacket setpoint temperature Tjsp, the faster the temperature transient, possibly reaching unsafe conditions if the deviation exceeds two standard deviations from the nominal value. The middle-right plot of Fig. 3 shows that the smaller the value of the jacket mass flow Fj, the slowest the transient during the heat-up phase, because less heat is transferred to the reactor. On the other hand, during the cooldown phase, less heat is removed from the reactor compared to the nominal case and the reactor temperature increases above the maximum values reached under nominal conditions, until sufficient heat has been transferred to the jacket flow to remove the heat produced by the exothermic reaction. An unsafe condition (system failure) is assumed to occur if the reactor temperature reaches 150 °C.

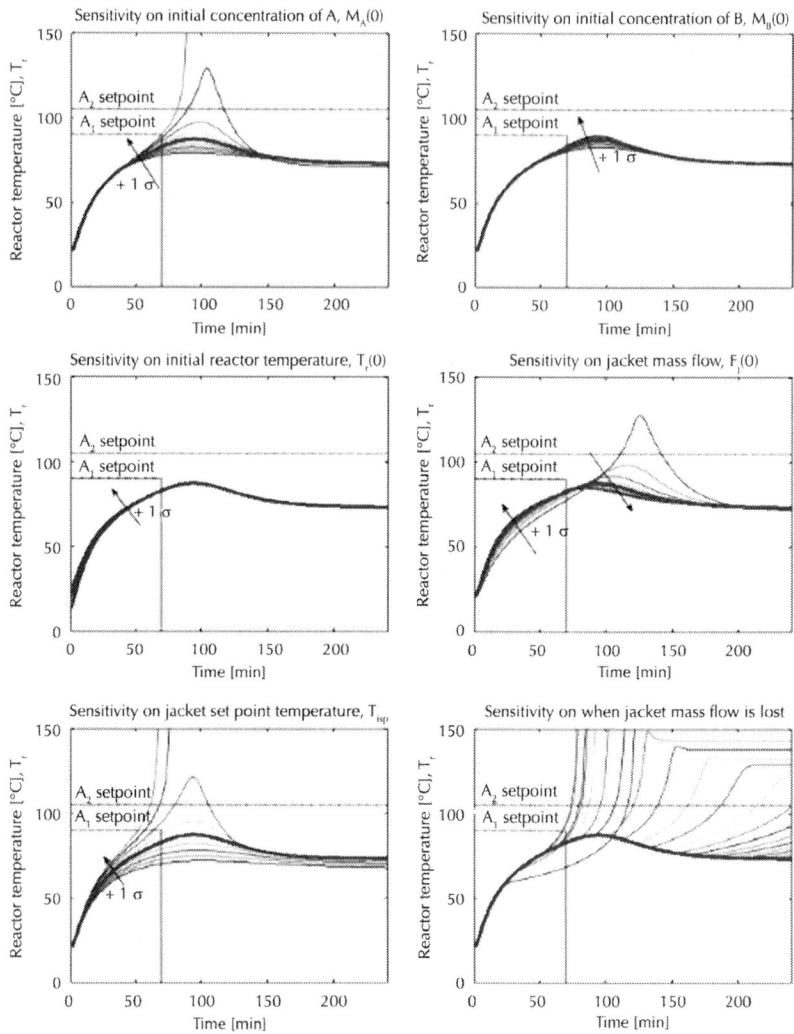

Figure 3: Behavior of the reactor temperature for different values of the process parameters and as a function of when the jacked flow is lost (bottom-right plot)—each behavior obtained by varying the parameter by one standard deviation or by progressively shifting by 10 min the time when the flow is lost. In bold: behavior obtained with nominal values.

Runaway conditions can also arise in case the system loses the thermal regulation provided by the water flow in the reactor jacket due, for example, to failures of the pumps that inject water into the jacket. This aspect is modeled by associating to the event of loss of flow rate Fj a constant occurrence rate of QF=1e−6 min^{-1}. As shown in the bottom-right plot of Fig. 3, the loss of flow leads to unsafe conditions if the reagent concentrations are sufficiently large.

Protective measures to avoid unsafe conditions are discussed next. Table 2 lists the assumed system and component failure rates and probabilities (except for the operator failure model which is reported in Table 3 and Table 4).

Table 2: Failure rates and probabilities assumed for the batch reactor model (see Table 3 and Table 4 for the operator action failure model)

Component/system	Failure rate/ probability
Alarm 1	1e−2 per demand
Alarm 2	1e−2 per demand
Jacket flow (total loss of jacket flow). Models effect of failure of the TCS	1e−6 min^{-1}
ECS failure to start	1e−2 per demand
ECS fails to run (results in total loss of jacket flow)	5e−7 min^{-1}

Table 3: Values of QD(t-t_A$_1$) [24] (applicable to response to A2 as well)

t-t_A1	QD(t-t_A1)
<1	1
<5	0.2
<10	0.1
<20	0.01
<30	0.001
<40	0.0005
<50	0.0002

Table 4: Values of QEXE(Tr)

Tr	QEXE1(Tr)-reduce Tjsp to 40 °C	QEXE2-align ECS
90	0.001	0.001
100	0.005	0.001
110	0.01	0.001
120	0.05	0.001
130	0.1	0.001
140	0.5	0.001
150	1	0.001

Two temperature alarms A1 and A2 are devised. The first alarm is activated if the reactor temperature reaches the value of $T_{A1}=90$ °C within the first 70 min, and the second alarm if it reaches $T_{A2}=105$ °C at any time during the batch reaction. The alarm activation conditions are shown for reference in Fig. 3. If the first alarm A1 is activated, the operator is instructed to reduce the setpoint temperature to 40 °C by acting on the TCS. In case of loss of jacket flow the operator is instructed to start up the ECS (acting on the TCS setpoint temperature would be ineffective because of the loss of flow). If the reactor temperature reaches 105 °C, the Emergency Cooling System (ECS) automatically actuates connecting the reactor to an additional heat sink at the constant temperature of 20 °C (in practice reducing Tjsp to 20 °C). Alarm A2 is also activated: the operator should manually start the ECS in case it failed to start automatically (for simplicity, the ESC failure to start is assumed fully recoverable).

The response of the operator includes: decision to act and actual execution, indicated in the following as events OPD and OPEXE, respectively. The probability that the operator does not take the decision to act, QD(t-t_A1), or QD(t-t_A2), is a function of the time elapsed from the alarm annunciation, t-t_A1, or t-t_A2. Thus, the time t at which the operator decides to act is a random variable.

Table 3 gives the probability distribution of no operator decision before t-t_A1 and t-t_A2 taken from the diagnosis model of THERP [24]. Once the operator has decided to act, say at time t, he may fail in the execution of the action with probability, QEXE1(Tr), dependent on the current temperature of the reactor (Table 4): the larger the temperature, the larger the mental stress felt by the operator and, accordingly, the failure probability. For the execution failure probability in response to alarm A2, QEXE2, a constant value of 1e−3 per demand is assumed.

In a previous paper [2], the authors have identified two types of scenarios induced by combinations of the initial values of reactor parameters. These are as follows.

Alarm A_1 bypassed. The initial values of reactor parameters induce a slow temperature increase in the initial heat up phase and a runaway reaction in a later phase. In this way, the activation condition for alarm A1 (reactor temperature of 90 °C within the first 70 min) is bypassed (Fig. 4: all pass below the alarm activation point in the figure). These scenarios are characterized by combinations of values of the jacket setpoint temperature larger than nominal with values of the jacket flow smaller than nominal.

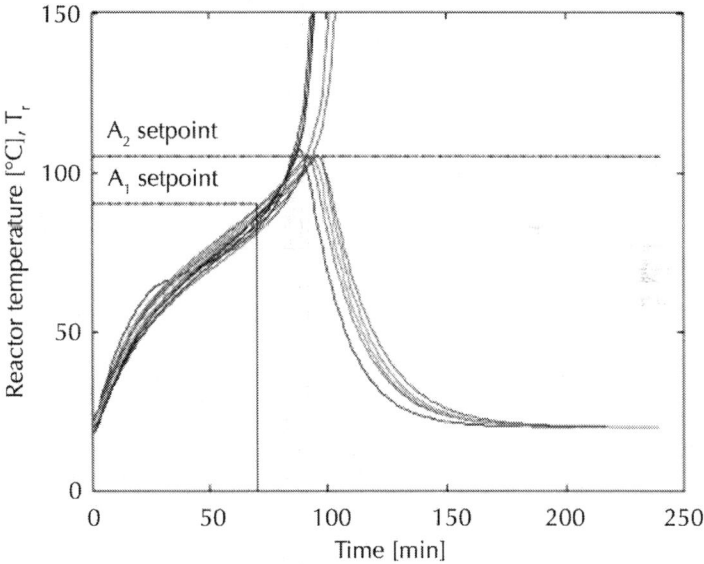

Figure 4: Behavior of the reactor temperature in scenarios example with alarm 1 bypassed.

Extreme deviations in the initial conditions. The deviations in the initial conditions are such to induce a very fast reactor dynamics that cannot be controlled before the reactor reaches the temperature of

150 °C, even if the operator or the ECS intervene correctly to reduce the jacket setpoint temperature. As an example, Fig. 5 shows how these uncontrollable transients develop as the value of the initial jacket setpoint temperature $T_{jsp}(0)$ increases.

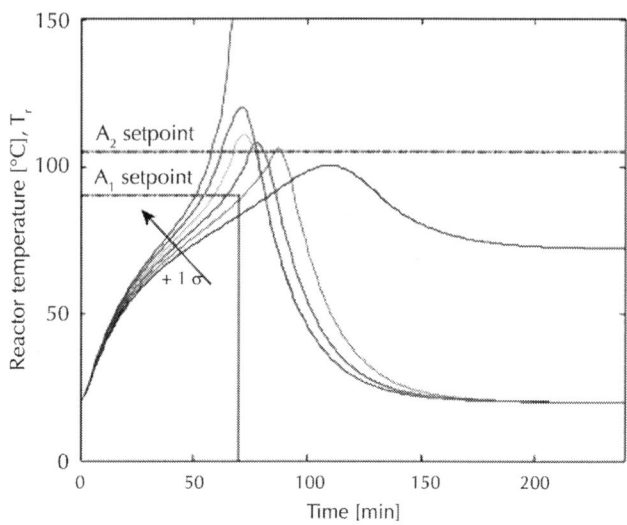

Figure 5: Development of the uncontrollable transients as the value of the initial jacket setpoint temperature $T_{jsp}(0)$ increases; each curve represents an increment of one standard deviation in Tjsp(0), from the case: MA(0)=13 kmol (nominal values +1σ); MA(0)=12 kmol (nominal value); $T_r(0)$=20 °C (nominal value); $F_j(0)$=0.248 m³/min (nominal values −2σ); $T_{jsp}(0)$=71 °C(nominal value).

PSA OF THE BATCH REACTOR

Scope of the PSA

The figure of risk considered in the study is the frequency of this temperature reaching 150 °C. The PSA, implemented on RiskSpectrum® software, includes two initiating events:

IE_DEV models the occurrence of transients due to deviations in the initial values of the process parameters such that the unsafe system state (reactor temperature exceeding 150 °C) may be reached (Table 1).

IE_TCS models the occurrence of transients initiated by a failure of the TCS. As noted in Section 2.2, the TCS failure is modeled as a total loss of jacket flow.

Event Trees and Fault Trees

The event tree for scenarios initiated by deviations in the initial process conditions, by initiating event IE-DEV, is shown in Fig. 6. The event tree was constructed based on the system description in Section 2 and the simulation runs in Fig. 3, Fig. 4, Fig. 5.

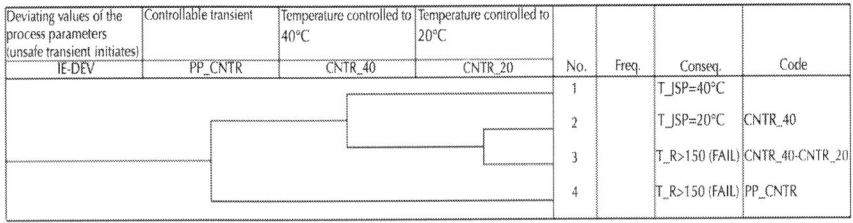

Figure 6: Event tree for initiating event IE_DEV (deviation in the process parameter values).

The event tree header PP_CNTR models whether the deviations of the initial values of the process parameters are "extreme". These are combinations of initial values of the process parameters that lead to very fast reactor dynamics that cannot be controlled before the reaction reaches 150 °C, even if the operator or the emergency system intervene correctly to reduce the jacket setpoint temperature. Failure of this top event implies that the transient is not controllable and directly leads to system failure (i.e. reactor temperature above 150 °C). Top event PP_CNTR is constituted by a single basic event (PP-EX).

The header CNTR_40 models the activation of alarm A1 and the response of the operator to reduce the jacket setpoint temperature to 40 °C. As shown in Fig. 7, failure of this top event occurs if alarm A1 fails or the operator fails to respond to the alarm, as well as when the alarm activation conditions are bypassed.

228   Chemical Process Safety

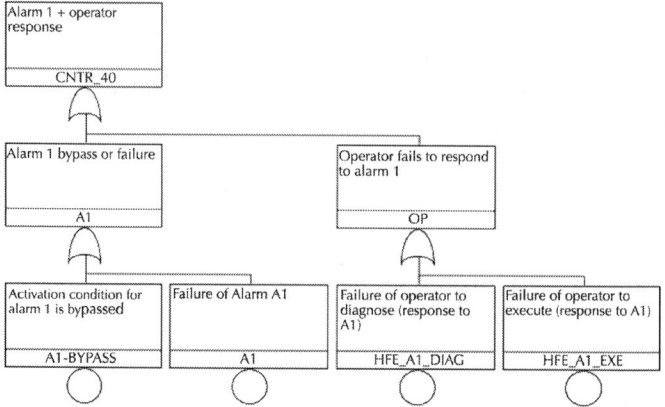

Figure 7: Fault tree for failure to control reactor temperature to 40 °C (CNTR_40).

The header CNTR_20 models the response of the ECS (automatic or manual start and successful run), as shown in Fig. 8. Failure of this top event leads to system failure (i.e. reactor temperature above 150 °C).

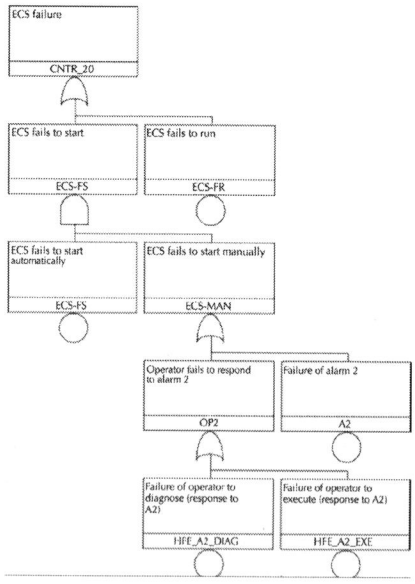

Figure 8: Fault tree for failure to control reactor temperature to 20 °C (CNTR_20).

For scenarios initiated by a failure of the Temperature Control System, representing a loss of jacket flow (initiating event IE-TCS), most of the transients bypass the activation conditions for alarm A1 (see the bottom-right plot in Fig. 3). Therefore, the corresponding event tree has only one top event, CNTR_20. The event tree developed for IE-TCS is not shown for brevity.

Quantification of the Initiating and Basic Event Probabilities

Table 5 gives the frequencies of the initiating events and the probabilities of the basic events included in the PSA model. Their quantification is discussed in the next sub-Section. As shown in Table 5, for simplicity, uncertainty in the reliability parameters is not considered—they would interact with the uncertainties in the process parameters and further complicate the system analysis. Indeed, it has to be mentioned that consideration of parameter uncertainty is fundamental in any PSA application, and could indeed lead to difference in the dynamic and traditional PSA results [4].

Table 5: List of the basic initiating events included in the model

ID	Description	Frequency/ probability used in the PSA	Modeled under top event
IE-DEV	Deviating values of the process parameters (unsafe transient initiates)	2.7e−2/day	IE-DEV
IE-TCS	Unsafe transients initiated by loss of jacket flow	9.0e−5/day	IE-TCS
PP-EX	Deviating values of the process parameters such that the transient is uncontrollable	1.1e−3	PP_CNTR

A1-BYPASS	Values of the process parameters such that the activation condition for alarm 1 is bypassed	1.0e−1	CNTR_40
A1	Alarm A1 failure	1.0e−2	CNTR_40
HFE_A1_DIAG	Failure of operator to diagnose (response to A1)	1.0e−1, analyst A	CNTR_40
		1.7e−1, analyst B	
HFE_A1_EXE	Failure of operator to execute (response to A1)	1.1e−3, analyst A	CNTR_40
		1.1e−3, analyst B	
ECS-FS	Emergency Cooling System fails to start	1.0e−2	CNTR_20
ECS-FR	Emergency Cooling System fails to run	2.50E−06	CNTR_20
A2	Alarm A2 failure	1.0e−2	CNTR_20
HFE_A2_DIAG	Failure of operator to diagnose (response to A2)	2.5e−1, analyst A	CNTR_20
		Guaranteed failed, analyst B	
HFE_A2_EXE	Failure of operator to execute (response to A2)	1.0e−3, analyst A	CNTR_20
		Guaranteed failed, analyst B	

Basic Events Determined by the Values of the Process Parameters

The occurrence of three events included in the PSA model is determined by the values of the five process parameters of Table 1. These events are: IE-DEV (which is an IE that represents values of the process parameters deviating from their nominal values to the extent that the unsafe system state, i.e. reactor temperature above 150 °C, may be reached), PP-EX (deviating values of the process parameters such that the transient is uncontrollable), A1-BYPASS (values of the process parameters such that the activation condition for alarm 1 is bypassed).

Quantifying the frequency/probability of these events entails quantifying the frequency/probability of the combinations of parameter values that lead to the events. The method of discrete probability distributions, often used in seismic PSA, is adopted [23]. The idea is

Conventional and Dynamic Safety Analysis: Comparison on a... 231

to approximate the continuous probability density function of each uncertain parameter with a discrete probability distribution, identified by a set of evaluation points xi and corresponding probabilities pi. Each point xi is the mean taken over each discretization interval identified by the pi›s. The intervals can be selected to explore the most interesting regions of the parameter values [23].

Table 6 shows the discretization cases analyzed to quantify events IE-DEV, PP-EX and A1-BYPASS. The discretizations are made on the values of the three parameters MA(0), Fj(0), Tjsp, to which the reactor temperature behavior is mostly sensitive (see Fig. 3), while the values of MB(0) and Tr(0) are kept at their nominal values. Different discretization cases need to be tried to check the stability of the estimates. The discretization cases differ for: the number of intervals/evaluation points (3 or 5 per parameter) and the size of the intervals (which determine which regions of the parameter values are explored, typically, tails vs. central region).

Table 6: Discretization cases for the evaluation of the probability of events which occur based on the values of the uncertain parameters

Discrete probability distributions	#Discretizations[b]	Probability of each evaluation interval[c]					Evaluation point of each interval[c,d]					
IE DEV	Unsafe reaction due to deviating process parameters						Event frequency/day					
Case 1	3	0.5	0.25	0.25			−0.8	0.3	1.3			1.6e−2
Case 2	5	0.5	0.125	0.125	0.125	0.125	−0.8	0.2	0.5	0.9	1.7	2.7e−2[a]
PP-EX	Deviating values of the process parameters such that the transient is uncontrollable						Probability (conditional on IE)					
Case 1	3	0.5	0.25	0.25			−0.8	0.3	1.3			−
Case 2	5	0.5	0.125	0.125	0.125	0.125	−0.8	0.2	0.5	0.9	1.7	−
Case 3	3	0.5	0.49	0.01			−0.8	0.76	2.7			3.7e−5
Case 4	3	0.5	0.499	0.001			−0.8	0.79	3.4			3.7e−5
Case 5	5	0.5	0.163	0.163	0.163	0.01	−0.8	0.2	0.7	1.4	2.7	6.3e−4
Case 6	5	0.5	0.47	0.01	0.01	0.01	−0.8	0.7	2.0	2.2	2.7	4.8e−4
Case 7	5	0.5	0.166	0.166	0.166	0.001	−0.8	0.2	0.7	1.5	3.7	1.1e−3[a]
Case 8	5	0.5	0.497	0.001	0.001	0.001	−0.8	0.8	2.8	3.0	3.4	2.2e−4
A1-BYPASS	Values of the process parameters such that activation condition for alarm 1 is bypassed						Probability (conditional on IE)					
Case 1	3	0.5	0.25	0.25			−0.8	0.3	1.3			−
Case 2	5	0.5	0.125	0.125	0.125	0.125	−0.8	0.2	0.5	0.9	1.7	7.2e−2
Case 3	5	0.5	0.163	0.163	0.163	0.01	−0.8	0.2	0.7	1.4	2.7	6.1e−2
Case 4	5	0.5	0.47	0.01	0.01	0.01	−0.8	0.7	2.0	2.2	2.7	1.0e−1[a]

[a] Value used in the PSA.

[b] The discretizations are made on the range of the three parameters MA(0), Fj(0), Tjsp. The number of model evaluations needed per case is $3^3=27$ or $5^3=125$.

[c] The values in the table refer to parameters MA(0) and Tjsp, for which the region of interest is towards values larger than nominal. For Fj(0), the values apply in revered order/sign so as to cover the region of values smaller than nominal.

[d] The evaluation points are given in standard deviations from the nominal value.

Two aspects are important to note from Table 6. First, not all discretization cases allow quantifying the events. For example, as reported in Table 6, in the two cases 1 and 2 for PP-EX the evaluation points do not deviate enough from the nominal values to cover values inducing uncontrollable conditions. Second, the obtained estimates can vary quite significantly, e.g. one order of magnitude in case of PP-EX in Table 6and it is not obvious which value is the most accurate (depends on number of intervals and how well the intervals cover the most "interesting" regions, which, of course, it is not known a priori).

The highest estimates from the discretization cases are then used in the PSA.

The question arises here of whether the simulation runs presented in Table 6 (likewise those reported inFig. 3, Fig. 4 and Fig. 5) allow covering all possible scenario variants. Of course, full completeness of the scenarios considered can never be assured with certainty. As in any industrial PSA, it is up to the analyst(s) to judge whether the space has been sufficiently covered. In the present application, the mentioned simulation runs (which amount to about 1000) give reasonable assurance on the completeness of the considered scenarios.

Operator Actions (HFE_A1_DIAG, HFE_A1_EXE, HFE_A2_DIAG)

The success criteria of operator actions are typically defined based on a bounding reference scenario. In order to show the impact of the choice of bounding scenarios, two hypothetical analysts are introduced, analysts A and B, who choose different scenarios (A and B). Analyst

A's approach is to select a scenario that yields a representative time window; in other words, if one were to sample the scenarios, the available time would often be comparable to this time window. Analyst B's approach is to select a scenario with the shortest time window for which the operator has a non-negligible change of success; he obtains a 6-minute time window.

Both analysts define a bounding scenario with a reactor temperature behavior with an inflection point around 40 min, where the alarm A1 setpoint is reached at approximately 50 min. If the operator or automatic system response fails, the reactor temperature continues to increase and system failure occurs at 70 min. Bounding scenario A is based on a deviation of the jacket setpoint temperature of three standard deviations, with all other parameters at their nominal values. It is the scenario in the bottom-left plot of Fig. 3 for which the runaway conditions establish at the smallest deviation of the value of the jacket setpoint temperature. Scenario B is at the onset of uncontrollable conditions shown in Fig. 5. It can be expected that scenario B is a more conservative choice (because it is closer to being uncontrollable), while scenario A is more representative of typical runaway scenarios.

Fig. 9 shows how the success criteria are derived. Each trajectory in the plots corresponds to a different time when the operator responds to alarms A1 and A2 (each branch corresponds to additional two minutes to the previous one). If scenario A is taken as reference, then 10 min is the maximum available time for the operators to respond to alarm A1 in order to avoid reaching a temperature of 150 °C (Fig. 9, left). If scenario B is taken as a reference, then this time is reduced to 6 min (Fig. 9, right). Based on the time reliability curve of Table 3, the result is a diagnosis failure probability (HFE_A1_DIAG) of 1.0e−1 and 1.7e−1, for analysts A and B respectively. Concerning the response to alarm A2, if scenario A is taken as reference, then the maximum time is 4 min (Fig. 9, left). If instead scenario B is taken as reference, then it can be seen that the operator has no possibility to control the reactor temperature in response to alarm 2 before it reaches 150 °C (Fig. 9, right). Analyst A would therefore quantify the probability of HFE_A2_DIAG as 2.5e−1, whereas analyst B would not credit this operator action. Note the probability of the operator failing to respond to alarm A2 (under analyst A›s assumptions) is assumed to be independent on whether the preceding failure of top event CNTR_40 is due to an operator failure or not. Methods for possibly treating this dependence can be found in Refs. [24] and [25].

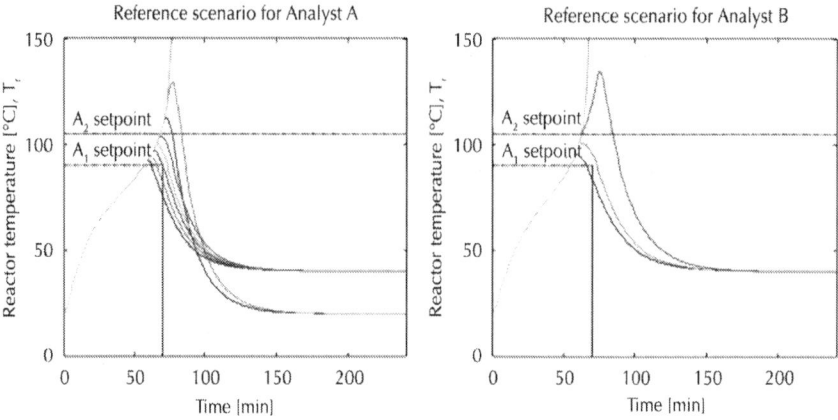

Figure 9: Definition of the success criteria for operator response: reference Scenario A (left) and reference scenario B (right)—each trajectory corresponds to progressively adding a delay of 2 min in the operator response.

Basic event HFE_A1_EXE (execution of the response to alarm 1) is quantified again based on Fig. 9, considering the value of the reactor temperature at the time when the operators have made the decision to reduce the setpoint temperature (Table 4). The values obtained are: 1.1e−3 by analyst A and 1.0e−3 by analyst B. The execution failure probability for the response to alarm 2 (HFE_A2_EXE) is independent of the reactor temperature (1.0e−3, Table 4).

Loss of Jacket Flow Events (IE-TCS, ECS-FR)

Concerning the initiating event IE-TCS, the loss of jacket flow results in an accident initiator only if it occurs when the reagent concentrations are sufficiently large. As shown in Fig. 3, this happens if the loss of flow occurs between 10 and 100 min from the reaction start. The frequency of event IE-TCS can be then quantified as

Fr(IE_TCS)=Fr(batch reaction process) Pr(loss of flow occurs within 10–100 min)==1/day 1e−6 [min^{-1}] (100−10) [min]=9e−5/day.

Similar considerations apply for the quantification of event ECS-FR. The failure of the ECS results in an unsafe transient only if the concentrations of the reagents are large enough at the time when the failure occurs. The probability of basic event ECS-FR was estimated at 2.5e−6/day.

PSA Results

The system failure frequency (the frequency of reactor temperature exceeding 150 °C) is quantified at 4.4e−5/day and 1.0e.5/day by analyst A and B, respectively (Table 7). The larger value obtained by analyst B is the result of the more conservative assumptions made by this analyst.

Table 7: PSA results: system failure frequency under the two analysts' assumptions

	Analyst A	Analyst B
System failure frequency	4.4e−5/day	1.0e−4/day

Table 8 and Fig. 10 compare the results in terms of basic event contribution to the system failure frequency (as measured by the Fussell–Vesely, FV, importance measure). Under both analysts' assumptions, scenarios initiated by deviations of the process parameters largely dominate over those initiated by the failure of the temperature control system (the contribution of IE-DEV and IE-TCS is 99% and 1%, respectively). The PSA results are also in agreement concerning the dominant basic events. A high importance group of events can be identified in Fig. 10 including, under both analysts' assumptions, the development of the so-called extreme scenarios (basic event PP-EX, see Section 3.2), the failure of the ECS to start (ECS-FS) after the setpoint for alarm A2 is reached, the failure of the operator to respond to alarms A1 and A2 (HFE_A1_DIAG and HFE_A2_DIAG) and the bypass of the activation conditions for alarm A1 (basic event IE-BYPASS).

Table 8: PSA results: Fussell–Vesely (FV) importance measures under the two analysts' assumptions

	ID	Description	Frequency/probability	FV, Analyst A	FV, Analyst B
1	IE-DEV	Deviating values of the process parameters (unsafe transient initiates)	2.7e−2/day	9.9e−1	9.9e−1

2	PP-EX	Deviating values of the process parameters such that the transient is uncontrollable	1.1e−3/day	6.8e−1	2.94e−1
3	ECS-FS	Emergency Cooling System fails to start	1.0e−2	3.2e−1	7.06e−1
4	HFE_A2_DIAG	Failure of operator to diagnose (response to A2)	2.5e−1, analyst A	3.1e−1	–
			Guaranteed failed, analyst B		
5	HFE_A1_DIAG	Failure of operator to diagnose	1.0e−1, analyst A	1.6e−1	4.54e−1
		(response to A1)	1.7e−1, analyst B		
6	A1-BYPASS	Values of the process parameters such that the activation condition for alarm 1 is bypassed	1.0e−1	1.6e−1	2.67e−1
7	A1	Alarm A1 failure	1.0e−2	1.6e−2	2.67e−2
8	A2	Alarm A2 failure	1.0e−2	1.3e−2	–
9	IE-TCS	Unsafe transients initiated by loss of jacket flow	9.0e−5	5.3e−3	8.90e−3
10	HFE_A1_EXE	Failure of operator to execute (response to A1)	1.0e−3	1.6e−3	2.67e−3
11	HFE_A2_EXE	Failure of operator to execute (response to A2)	1.0e−3, analyst A Guaranteed failed, analyst B	1.3e−3	–
12	ECS-FR	Emergency Cooling System fails to run	2.5e−6	3.1e−4	1.77e−4

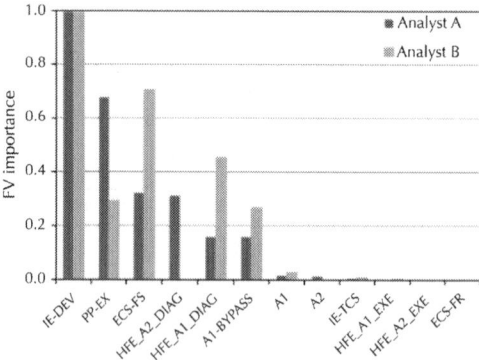

Figure 10: PSA results: Fussell–Vesely (FV) importance measures under the two analysts' assumptions.

Indeed, Table 8 and Fig. 10 show differences in the relative contribution of the events as assessed by analysts A and B. Recall from Section 3.3.2 that the scenario chosen by analyst B is more challenging than that chosen by analyst A: a faster operator response to alarm A1 is required and no operator response to alarm A2 is credited. For analyst A, the top contribution to risk comes from the extreme condition scenarios (68% for PP-EX); then, from two failures of the measures activated at the setpoint of alarm A2 (failure of the ECS to automatically start, ECS-FS with 32%, and failure of the operator to respond timely to alarm A2, HFE_A2_DIAG with 31%); and, finally, from two other events related to the first protection barrier contribute around 16% (failure of the operator to respond timely to alarm A1 and the conditions of activation of alarm A1 being bypassed). Under the assumptions by analyst B, the contribution by the failure of the protective measures (i.e. events ECS-FS, HFE_A2_DIAG; HFE_A1_DIAG; A1-BYPASS) is more important than that from the extreme condition scenarios (Table 8 and Fig. 10). This is the result of the reference scenario chosen by analyst B: the more conservative choice (a more challenging transient to control) increases the contribution to the risk of the affected sequences (i.e. those modeling protective measures, as opposed to the extreme condition scenarios).

Similar considerations can be made comparing the PSA results in terms of identified MCS (Table 9). The frequency of uncontrollable transients (MCS #1 of analyst A and MCS #2 of analyst B) does not depend on the assumptions made on the operator response modeling therefore is the same for analysts A and B. The frequencies of the other MCSs reflect the differences in the assumptions by the two analysts. For example, for MCS #2 of analyst A and MCS #1 of analyst B, the difference in frequency is due to the different probability of diagnosis failures used (1e−1 for analyst A and 1.7e−1 for analyst B) and to the fact that analyst B cannot credit the operator response to alarm 2 (in turn, a consequence of the chosen bounding scenario).

Table 9: PSA results: minimal cuts sets with frequency above 2.5e-7/day under the two analysts' assumptions

Analyst A				Analyst B			
#	Frequency [/day]	%	Minimal cut sets	#	Frequency [/day]	%	Minimal cut sets

1	3.0e−5	68.1	IE-DEV	1	4.6e−5	45.4	IE-DEV
			PP-EX				HFE_A1_DIAG
							ECS-FS
2	6.8e−6	15.5	IE-DEV	2	3.0e−5	29.4	IE-DEV
			HFE_A1_DIAG				PP-EX
			ECS-FS				
			HFE_A2_DIAG				
3	6.8e−6	15.5	IE-DEV	3	2.7e−5	26.7	IE-DEV
			A1-BYPASS				A1-BYPASS
			ECS-FS				ECS-FS
			HFE_A2_DIAG				
4	6.8e−7	1.6	IE-DEV	4	2.7e−6	2.7	IE-DEV
			A1				A1
			ECS-FS				ECS-FS
			HFE_A2_DIAG				
5	2.7e−7	0.6	IE-DEV	5	9.0e−7	0.9	IE-TCS
			HFE_A1_DIAG				ECS-FS
			ECS-FS				
			A2				
6	2.7e−7	0.6	IE-DEV	6	2.7e−7	0.3	IE-DEV
			A1-BYPASS				HFE_A1_EXE
			ECS-FS				ECS-FS
			A2				

MONTE CARLO ANALYSIS

The Monte Carlo Approach for Dynamic Analysis

Monte Carlo (MC) simulation for system safety and reliability analysis entails generating many system life histories (referred to as "MC trials" in what follows) by sampling the transition times and states of the system components and the outcomes of operators actions [26].

The dynamic features of the system evolution are embedded within the stochastic simulation by introducing the physical models of the process evolution (i.e. Eqs. (1), (2), (3), (4), (5) and (6) for the present application) [3]. At the beginning of each simulation, the values of the process parameters are sampled from Gaussian distributions with the characteristics reported in Table 1.

During each MC trial, a record is kept of all the events that have occurred (e.g. whether the operator has responded to an alarm, whether the ECS has successfully started, and the like), independently of the time or sequence of occurrence. During the simulation, a record archive is maintained. At the end of each MC trial, the corresponding record is compared to those already present in the archive:

- if the record matches one record already present in the archive, the counter associated to that record is incremented by 1.
- if the record does not match any of the records already present in the archive, then the new record is added to the archive.

Note that the last element of the record indicates whether the trial has concluded with a system success (reactor temperature below 150 °C) or failure (reactor temperature above 150 °C). This allows distinguishing between records characterized by the same set of events, but reaching two different final states, for example because of differences in the timing when (or in the order in which) these events occur.

The MC Results

The system failure frequency is estimated as $2.7e{-}5 \pm 1.2e{-}6$/day (based on $2e7$ MC trials).

The contribution of the events to the system failure is assessed on the basis of the record archive at the end of the simulation (reported in Table 10). In particular, the FV measure is computed following its definition in the traditional PSA framework, but applied to the MC records instead of the MCSs. Then, the definition of the FV importance of an event is: the ratio of the frequency of the records containing that event, to the total system failure frequency (the implication of using MC records instead of the MCSs will be discussed later in this Section). The MC-estimated measures are reported in Fig. 11. First, the results show that the risk is dominated by scenarios initiated by deviations in the process parameters (IE-DEV and IE-TCS contribute 99% and 1%, respectively). Then, the top contributing events relate to the development of the extreme scenarios, the failure of the ECS to start, the failure of the operator to respond to the alarms and the bypass of the activation conditions for alarm A1.

Table 10: System failure records at the end of the simulation (records with more than 5 histories, i.e. with estimated frequency above 2.5e-7±1.1e-7/day)

Record#	TCS	Alarm 1	Operator response to alarm 1		ECS			Operator response to alarm 2		MC estimate [/d]	
			Diagnosis	Execution	Start	Run	Alarm 2	Diagnosis	Execution	Mean	Standard deviation
1	S	by-passed	–	–	F	–	S	Not made	–	5.0e–6	5.0e–7
2	S	by-passed	–	–	S	S	–	–	–	5.0e–6	5.0e–7
3	S	by-passed	–	–	F	–	S	Made	S	3.8e–6	4.4e–7
4	S	S	Not made	–	F	–	S	Made	S	3.8e–6	4.4e–7
5	S	S	Not made	–	F	–	S	Not made	–	3.5e–6	4.2e–7
6	S	S	Not made	–	S	S	–	–	–	3.3e–6	4.1e–7
7	S	F	–	–	S	S	–	–	–	4.5e–7	1.5e–7
8	S	by-passed	–	–	F	–	F	–	–	4.5e–7	1.5e–7
9	S	F	–	–	F	–	S	Made	S	3.5e–7	1.3e–7
10	S	S	Made	S	S	S	–	–	–	3.0e–7	1.2e–7
11	S	F	–	–	F	–	S	Not made	–	3.0e–7	1.2e–7
12	S	S	Not made	–	F	–	F	–	–	2.5e–7	1.1e–7
13	F	by-passed	–	–	F	–	S	Made	S	2.5e–7	1.1e–7

Event coding: "S" success; "F" failure; "–" event not possible (e.g. diagnosis not possible if alarm has failed or has been bypassed).

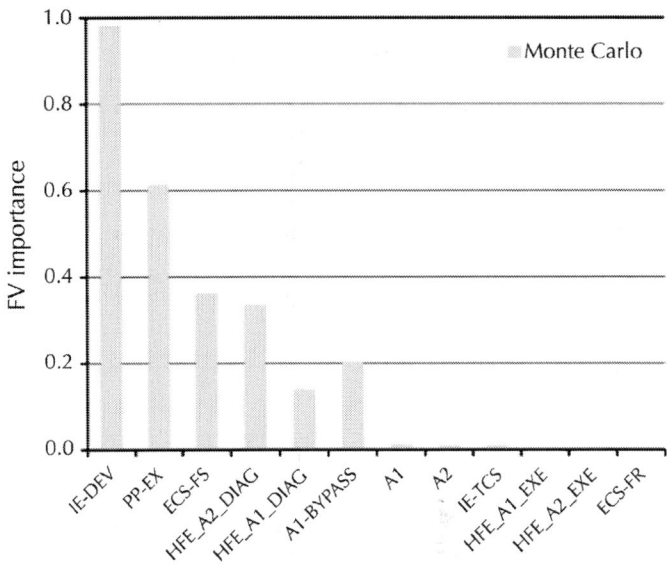

Figure 11: MC results: Fussell–Vesely (FV) importance measures.

A first remark relates the quantification of the contribution of the extreme condition scenarios in the MC framework: these scenarios have to be recognized among the archived records in Table 10. According to their definition in Section 3.2, these are scenarios in which the initial conditions are such to induce a very fast reactor dynamics that cannot be controlled before the reactor reached the temperature of 150 °C, even if the operator or the ECS intervene correctly. Therefore, in Table 10, these scenarios are represented by records in which the protective measures (operator or ECS) respond as designed, but the system nevertheless fails (in Table 10, those records with 'S' in ECS run or 'S' in operator response to alarm 2-execution, i.e. #2, 3, 4, 6, 7, 9, and 10). The FV contribution of these scenarios is then the ratio of the frequency of the corresponding records, to the system failure frequency.

Note also that Table 11 and Fig. 11, related to the MC results, use the same event identifier as the corresponding PSA basic event. This is to ease the MC-PSA comparison in the next Section, but an important difference should be noted. In the PSA framework, basic event occurrences are connected with corresponding success criteria (i.e. operator diagnosis within a given time frame). In the MC framework, the event is recorded if it has occurred, independently of whether the success criteria are met (e.g. the operator has diagnosed the transient, even though the diagnosis is reached too late).

Table 11: MC results: Fussell–Vesely (FV) importance measures

	ID	Description	FV
1	IE-DEV	Deviating values of the process parameters (unsafe transient initiates)	9.9e−1
2	PP-EX	Deviating values of the process parameters such that the transient is uncontrollable	6.1e−1
3	ECS-FS	Emergency Cooling System fails to start	3.6e−1
4	HFE_A2_DIAG	Failure of operator to diagnose (response to A2)	3.4e−1
5	HFE_A1_DIAG	Failure of operator to diagnose (response to A1)	1.4e−1
6	A1-BYPASS	Values of the process parameters such that the activation condition for alarm 1 is bypassed	2.0e−1
7	A1	Alarm A1 failure	1.1e−2
8	A2	Alarm A2 failure	9.3e−3
9	IE-TCS	Unsafe transients initiated by loss of jacket flow	9.3e−3
10	HFE_A1_EXE	Failure of operator to execute (response to A1)	−[a]
11	HFE_A2_EXE	Failure of operator to execute (response to A2)	−[a]
12	ECS-FR	Emergency Cooling System fails to run	−[a]

[a]These events never occurred in the MC simulation with 2e7 trials; they do not appear in any MC record and their contribution cannot be estimated with the mentioned procedure.

Related to the above remark is the fact that the MC records may include non-minimal accident sequences, i.e. including failures that are not "needed" to lead to the system failure, or in other words, the system failure is reached independently on the outcome of these events. Consider, for example, record #1 in Table 10, in which the operator failed to respond to alarm A2. In some sequences archived under record #1, diagnosis success would have led to the system success: in these sequences the diagnosis failure actually contributed to the system failure. In some other sequences (still archived as record #1), even a diagnosis success would not have prevented the system failure, because the reactor transient was already uncontrollable, regardless of the operator diagnosing it or not: in these sequences the diagnosis failure did not contribute to the system failure, although the diagnosis failure event is recorded. Such sequences should actually be recognized as extreme deviation scenarios, but the two types of sequences are achieved under the same record #1 (because the same events occur).

Then, the effect of the presence of non-minimal sequences is that the frequency of the records corresponding to the extreme condition scenarios (records #2, 3, 4, 6, 7, 9, 10) is underestimated, while that of the records in which the protective measures are actually effective is overestimated (#1, 5, 8, 11, 12, 13). Consequently, if MC records are used to calculate the event contribution to the system failure (the FV importance measure), the calculated contribution of event PP-EX is underestimated, while that of the other events is overestimated, because it includes "not needed" failures.

The fact that non-minimal sequences are generated is an inherent characteristic of the MC simulation approach, because events are sampled as they would occur in a real system, independently of whether they actually cause the system failure. Then, the issue is whether the effect on the FV calculation is such to distort the risk contributor profile of Fig. 11. This cannot be known a priori, because it depends on the characteristics of the analyzed system (ultimately, on the likelihood of the non-minimal sequences with respect to the minimal ones). In the present paper, which focuses on a comparison of PSA and MC results and not on the MC results in absolute terms, the PSA-MC comparison has been made bearing the presence of non-minimal sequences in mind, and considering their effect a posteriori, in relation to the PSA results.

COMPARISON OF THE RESULTS

The PSA and MC results are compared both in terms of overall risk figure (system failure frequency) and risk contributors.

For the risk figure, the MC estimate represents the "true" value to which the PSA result is compared: the closer the PSA result to the MC estimate, the more accurate the PSA results are. Under the assumptions of analyst A, the PSA result of 4.4e−5/day is about a factor of 1.5 larger than that from the MC dynamic analysis, while under the assumptions of analyst B, the result of 1.0e−4/day is about a factor of 4 larger. The results are consistent with the choice of the reference scenarios for the derivation of the success criteria (Section 3.3.2). The scenario chosen by analyst A is more representative of the typical runaway scenarios, and, consequently, the PSA result is closer to that from the MC simulation. The scenario chosen by analyst B is a more conservative choice (more challenging scenario to control), therefore leading to more conservative results.

For the risk contributors, Fig. 12 compares the FV measures (these are the same values as reported in Table 8, Table 11, Fig. 10 and Fig. 11). The PSA and MC approaches reach the same conclusion in terms of which events are the dominant risk contributors. As already discussed in the earlier sections, these are connected with the presence of "extreme" condition scenarios, the failure of the ECS to start, the failure of the operator to respond to the alarms A1 and A2 and the bypass of the activation conditions for alarm A1.

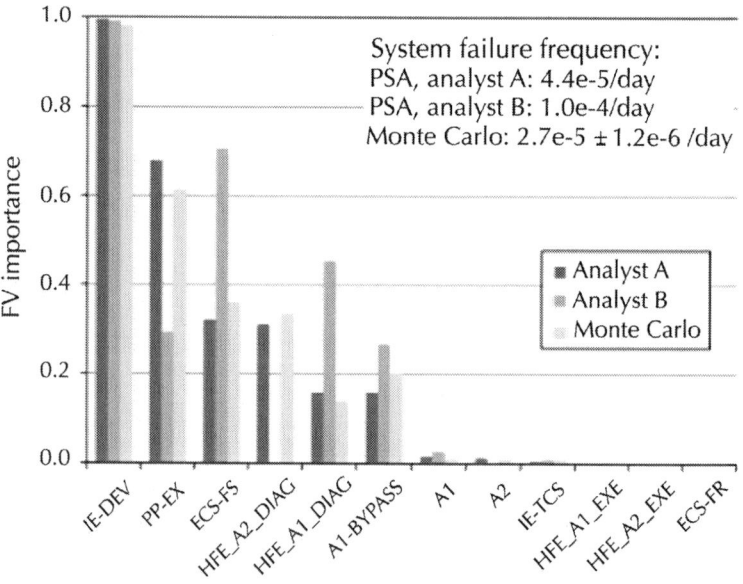

Figure 12: Comparison of MC and PSA results in terms of risk contributors (Fussell–Vesely, FV, importance).

Fig. 12 shows also a close match of the MC and the PSA results obtained with analyst A's assumptions. This suggests that the bounding scenario chosen by analyst A indeed represents well the behavior of transients leading to unsafe plant conditions. Also, the success criteria derived for the operator response to alarms represent well the conditions under which operators are called to act during these transients. Furthermore, the figure shows that the risk profile from the MC-derived FV measures is not significantly distorted by the non-minimal sequences.

The fact that the PSA results (under analyst A's assumptions) are in line with those from the MC can be confirmed by comparing the frequencies of the MC records in Table 10 and of the MCS in Table 9. As an example, record #1 in Table 10 represents the following failure sequence: an unsafe transient is initiated due to a certain combination of the process parameter values; the TCS has functioned through the whole batch duration; the activation conditions for alarm 1 have been bypassed; operators did not intervene (because of alarm not being activated); the ECS has failed to start automatically; and the operators have failed to timely diagnose the need for manual ECS start-up. Indeed, this sequence is represented by the following MCS found by analyst A: IE-DEV; A1-BYPASS; ECS-FS; HFE_A2_DIAG. The MCS frequency of 6.8e−6/day is quite close to that of the MC record 5.0e−6/day. Similarly, comparing the frequency of extreme deviation scenarios, the PSA value of 3.0e−5/day is in agreement with the MC estimate of 1.8e−5/day (obtained by summing the MC estimates of records #2, 3, 4, 6, 7, 9, 10).

Fig. 12 also shows that the relative importances obtained with analyst B's assumptions are quite different from those from the MC results (as well as analyst A's results). As already said in Section 3.4, this is the result of the conservative scenario chosen by analyst B for the success criteria definition: the more conservative choice (a more challenging transient to control) increases the contribution of the affected sequences (i.e. those modeling protective measures, as opposed to the extreme condition scenarios).

Note finally that importance measures provide insights in the system risk profile and are typically used to suggest measures to improve the system safety. If the ranking produced by the measures is not accurate (as it is the case for those produced under analyst B assumptions), the measures may be sub-optimal. For example, the ranking obtained by analyst B suggests improvements should be devoted to reduce the probability of failure of the ECS to start. However, as the MC results suggest and analyst A correctly identifies, more effective measures (cost and practicality considerations put aside) could be put in place to make sure initial parameters do not lead to uncontrollable scenarios.

DISCUSSION AND CONCLUSIONS

The paper has compared the results from a conventional PSA with those from a dynamic methodology (MC simulation) in terms of both the predicted risk figure and most contributing accident sequences. The comparison is made on a chemical batch reactor from the literature, processing a highly exothermic reaction.

As a general remark, the PSA and MC results appear quite close. In terms of predicted risk figure, the PSA value is about a factor of 1.5 larger than the MC result, or a factor of 4 if more conservative assumptions in the PSA analysis are made. The PSA identified all the top contributing events as they result from the MC simulation; the top 6 and dominant contributors are identical for MC, PSA A, and PSA B. Still, some differences were found in the relative importance of these events, depending on the PSA reference accident scenarios taken for the success criteria definition. These ranking differences suggest different risk profiles and could lead to the selection of potentially sub-optimal safety improvement measures. This is the case, in the present application, in presence of the two different analysts' results: if the MC results were not available, a decision maker would most likely rely on the results from PSA B, which are believed to be more conservative. While the predicted risk figure from PSA B is indeed conservative, the corresponding risk profile (from the FV importance measure) is less accurate than that from PSA A and decisions solely based on the importance measure ranking would be sub-optimal.

A generalized conclusion on the adequacy of conventional methods in treating dynamic systems is out of the scope of the present paper. The system analyzed is relatively complex in terms of process variables behavior, but certainly not in terms of the number of components/systems and operating crew interactions. In addition, the present study does not address digital I&C systems, the treatment of which is believed to challenge the traditional methods. From the present case study, it can be concluded that, provided that the success criteria are appropriately defined, conventional and dynamic methodologies produced consistent results. Yet, one important comment needs to be made concerning the conventional PSA analysis made in this paper. The PSA analysis has been based on many system simulation runs (at least 1260 system simulation runs were performed for this small system). A comparably exhaustive characterization of the scenario space would require a number of runs that may not be feasible in industrial-scale applications. A reduced set of simulation runs may not have allowed the deep understanding of the system evolution variants obtained in the present PSA analysis. Indeed, the present application has shown that fewer simulation runs (from the discretization cases for the uncertain input process parameters) would have led to much different probability and frequency estimates for the basic events and to the omission of important accident scenario variants (extreme deviations in the initial process parameter values and bypass of Alarm A1), which were instead directly identifiable by the dynamic MC simulation.

ACKNOWLEDGMENTS

Parts of this work were funded by the Swiss Federal Nuclear Safety Inspectorate (ENSI), under DIS-Vertrag nr. 82610. The views expressed in this article are solely those of the authors. The authors would like to thank Dr. Nathan Siu and Prof. Enrico Zio for their comments contributing to inspire this work.

REFERENCES

1. Siu N. Risk assessment for dynamic systems: an overview. Reliability Engineering and System Safety 1994;43:43–73.
2. Podofillini L, Mercurio D, Dang VN, Zio E. Dynamic safety assessment: scenario identification via a fuzzy clustering approach. Reliability Engineering and System Safety 2010;95:534–49.
3. Marseguerra M, Zio E. Monte Carlo approach to PSA for dynamic process systems. Reliability Engineering and System Safety 1996;52:227–41.
4. Labeau PE, Smidts C, Swaminathan S. Dynamic reliability: towards an integrated platform for probabilistic risk assessment. Reliability Engineering and System Safety 2000;68:219–54.
5. Sheng KS, Mosleh A. The development and application of the accident dynamic simulator for dynamic probabilistic risk assessment of nuclear power plants. Reliability Engineering and System Safety 1996;52:297–314.

6. Hakobyan A, Aldemir T, Denning R, Dunagan S, Kunsman D, Rutt B, et al. Dynamic generation of accident progression event trees. Nuclear Engineering and Design 2008;238:3457–67.
7. Kopustinkas V, Augutis J, Rimkevicius S. Dynamic reliability and risk assessmenent of the accident localization system of the Ignalina NPP RBMK-1500 reactor. Reliability Engineering and System Safety 2005;87: 77–87.
8. Acosta C. Dynamic event tree for accident sequence analysis. PhD thesis. Massachusetts Institute of Technology; June 1991.
9. Chang YHJ and Mosleh A. Cognitive modeling and dynamic probabilistic simulation of operating crew response to complex system accidents (ADSIDACrew). College Park, Maryland: Center for Technology Risk Studies, University of Maryland; 1999.
10. Metzroth K, Denning R, Smidts C, Aldemir T. Incorporation of a human reliability model into the ADAPT PRA methodology. 2008.
11. Mercurio D, Dang VN. Use of dynamic event trees to model variability in crew time performance. 2010.
12. Kloos M, Peschke J. Consideration of human actions in combination with the probabilistic dynamics method Monte Carlo dynamic event tree. Journal of Risk and Reliability 2008;222:303–13.
13. Aldemir T, Stovsky MP, Kirschenbaum J, Mandelli D, Bucci P, Mangan LA, et al. Dynamic reliability modeling of digital instrumentation and control systems for nuclear reactor probabilistic risk assessments (NUREG/CR-6942). Prepared for US nuclear regulatory commission; October 2007.
14. Yau M, Apostolakis G, Guarro S. The use of prime implicants in dependability analysis of software controlled systems. Reliability Engineering and System Safety 1998;62:23–32.
15. Pasman HJ, Jung S, Prem K, Rogers WJ, Yang X. Is risk analysis a useful tool for improving process safety? Journal of Loss Prevention in the Process Industries 2009;22:769–77.
16. Yang X, Sam Mannan M. The development and application of dynamic operational risk assessment in oil/gas and chemical process industry. Reliability Engineering and System Safety 2010;95:806–15.

17. Aneziris ON, Papazoglou IA. Fast Markovian method for dynamic safety analysis of process plants. Journal of Loss Prevention in the Process Industries 2004;17:1–8.
18. Rizal D, Tani S, Nishiyama K, Suzuki KA. Safety and reliability analysis in a polyvinyl chloride batch process using dynamic simulator-case study: Loss of containment incident. Journal of Hazardous Materials 2006;137:1309–20.
19. Li S, Bahroun S, Valentin C, Jallut C, De Panthou F. Dynamic model based safety analysis of a three-phase catalytic slurry intensified continuous reactor. Journal of Loss of Prevention in the Process Industry 2010;23: 437–45.
20. Lou HH, Chandrasekaran J, Smith RA. Large-scale dynamic simulation for security assessment of an ethylene oxide manufacturing process. Computers and Chemical Engineering 2006;30:1102–18.
21. Hauptmanns U. Different sets of reliability data and success criteria in a probabilistic safety assessment for a plant producing nitroglycol. Journal of Hazardous Materials 2009;162:1322–9.
22. Aropornwickanop A, Kittisupakon P, Mujtaba IM. On-line dynamic optimization and control strategy for improving the performance of batch reactors. Chemical Engineering and Processing 2005;44:101–14.
23. Kaplan S. On the method of discrete probability distributions in risk and reliability calculations—application to seismic risk assessment. Risk Analysis 1981;1:189–96.
24. Swain AD, Guttman HE. Handbook of human reliability analysis with emphasis on nuclear power plant applications, 78. U.S. Nuclear Regulatory Commission, NUREG/CR-12; 1983.
25. Cepin M. DEPEND-HRA—a method for consideration of dependency in human reliability analysis. Reliability Engineering and System Safety 2008;93: 1452–60.
26. Marseguerra M, Zio E. Basics of the Monte Carlo method with application to system reliability. LiLoLe—Verlag GmbH (Publ. Co. Ltd.); 2002.

Citations

CHAPTER 1

László Dobos, András Király, and János Abonyi, "Economic-Oriented Stochastic Optimization in Advanced Process Control of Chemical Processes," The Scientific World Journal, vol. 2012, Article ID 801602, 10 pages, 2012. doi:10.1100/2012/801602.

CHAPTER 2

Paula Saavalainen, Satish Kabra, Esa Turpeinen, et al., "Sustainability Assessment of Chemical Processes: Evaluation of Three Synthesis Routes of DMC," Journal of Chemistry, vol. 2015, Article ID 402315, 12 pages, 2015. doi:10.1155/2015/402315.

CHAPTER 3

Kritzer, P., Döring, H. and Emermacher, B. (2014) Improved Safety for Automotive Lithium Batteries: An Innovative Approach to include an Emergency Cooling Element. Advances in Chemical Engineering and Science, 4, 197-207. doi: 10.4236/aces.2014.42023.

CHAPTER 4

N. López-Castillo, A. Rojas-Rodríguez, B. Porta and M. Cruz-Gómez, "Process for the Obtention of Coumaric Acid from Coumarin: Analysis of the Reaction Conditions," Advances in Chemical Engineering and Science, Vol. 3 No. 3, 2013, pp. 195-201. doi: 10.4236/aces.2013.33025.

CHAPTER 5

Ofélia de Queiroz F. Araújo, José Luiz de Medeiros and Rita Maria B. Alves (2014). CO_2 Utilization: A Process Systems Engineering Vision, CO_2 Sequestration and Valorization, Mr. Victor Esteves (Ed.), ISBN: 978-953-51-1225-9, InTech, DOI: 10.5772/57560.

CHAPTER 6

John M. Woodley, Michael Breuer, Daniel Mink, A future perspective on the role of industrial biotechnology for chemicals production, Chemical Engineering Research and Design, Volume 91, Issue 10, October 2013, Pages 2029-2036, ISSN 0263-8762, http://dx.doi.org/10.1016/j.cherd.2013.06.023.

CHAPTER 7

Taha Benikhlef, Djamel Benazzouz, Smail Adjerid, Kazimierz Lebecki, Safety analysis approach based on thermodynamic and chemical reactions modelling, Journal of Loss Prevention in the Process

Industries, Volume 25, Issue 3, May 2012, Pages 494-504, ISSN 0950-4230, http://dx.doi.org/10.1016/j.jlp.2011.12.006.

CHAPTER 8

L. Podofillini, V.N. Dang, Conventional and dynamic safety analysis: Comparison on a chemical batch reactor, Reliability Engineering & System Safety, Volume 106, October 2012, Pages 146-159, ISSN 0951-8320, http://dx.doi.org/10.1016/j.ress.2012.04.010.

Index

A
Advanced Process Control (APC) 4
Azobisisobutyronitrile (AIBN) 19

B
Battery management system (BMS) 58

C
Carbon dioxide 53
Carbon emitter 88
Carbon geological storage (CGS) 91
Chemical equilibrium (ChE) 101
Chemical Process Systems Engineering (cPSE) 91
Chemical safety assessment (CSA) 31

D
Deoxyribonucleic acid (DNA) 158
Dimethyl carbonate (DMC) 29, 50
Dynamic event trees (DET) 198

E
Electric Vehicles 68
Environmental impact assessment (EIA) 29
Environmental, social, and governance (ESG) 29
Enzyme inhibition, hypotoxicity

74
European Federation of Chemical Engineering (EFCE) 158

F

Fischer-Tropsch (FT) 89, 124
Fussell–Vesely (FV) 235, 236, 241, 242

G

Generalized Pattern Search (GPS) 12
Global Reporting Initiative (GRI) 29
Global warming potential (GWP) 55

H

Hybrid Electric Vehicles 54, 68

I

Increased energy 87
Individual Probabilistic Constraint problem (IPC) 7
In situ product removal (ISPR) 167

L

Life-cycle analysis (LCA) 163
Life Cycle Assessment (LCA) 29

M

Mesh Adaptive Direct Search (MADS) 3, 10, 12, 23
Methyl-isocyanate (MIC) 182
Mobile air conditioning 54, 59, 70
Mobile air conditioning (MAC) 53, 54, 55, 56, 59, 70

Model Predictive Controlled (MPC) 4
Model Predictive Controllers (MPCs) 8
Monte Carlo (MC) 238
Monte Carlo Simulation (MCS) 11
Multicriteria assessment (MCA) 41, 46

N

Number average molecular weight (NAMW) 19

O

Operator Training Systems (OTS) 24

P

Plug-in Hybrid Electric Vehicles (PHEVs) 54
Pressure–volume (PV) 192
Probabilistic Safety Assessment (PSA) 215, 216

Q

Quantitative Risk Assessment (QRA) 215

R

Reactive Vapor-Liquid Equilibrium (RVLE) 100
Recombinant DNA (rDNA) 161
Reverse water gas shift (RWGS) 89

S

Safety data sheets (SDS) 31
Solid-liquid extraction system 76

State of charge 65
State of Charge 68
statistical process control (SPC) 2
Supercritical fluid (SCF) 96
Supercritical fluid zone (SCF) 94

T

Temperature Control System (TCS) 220
Triple bottom line\" (TBL) 28
Triple Point (TP) 93